FLORA OF THE GUIANAS

Edited by

M.J. JANSEN-JACOBS

Series A: Phanerogams

Fascicle 25

105a. EREMOLEPIDACEAE
&
105b. LORANTHACEAE
&
106. VISCACEAE
(J. Kuijt)

2007
Royal Botanic Gardens, Kew

Contents

INTRODUCTION AND KEY TO THE GUIANAN MISTLETOE FAMILIES

by

JOB KUIJT[1]

The basically Euro-centered view of all mistletoes contained in a single family, LORANTHACEAE, proved to be unacceptable by the second half of the 20th Century. All specialists since the 1960's have been in agreement that two major natural taxa exist, viz., LORANTHACEAE s.s. and VISCACEAE (Barlow 1964, Kuijt 1968). The major distinction between the two families is the presence of a calyculus (usually interpreted as a reduced calyx) in LORANTHACEAE only, the flower thus being dichlamydeous in contrast to the monochlamydeous Viscacean flower with a single floral whorl. In addition, there are embryological, chromosomal, and other distinctions (Kuijt 1969).

It appears that LORANTHACEAE s.s. are of southern origin while VISCACEAE probably originated in the Northern Hemisphere (Barlow & Wiens 1971, Wiens & Barlow 1971). The great majority of LORANTHACEAE (nearly all the Old World genera) are predominantly bird-pollinated and have large, colorful flowers, but no VISCACEAE appear to be.

Awareness of the exsistence of other groups of Santalalean arboreal parasites has further complicated the taxonomic conception of "mistletoe". In southeastern Asia, several groups of SANTALACEAE parasitize branches of woody hosts; especially striking is the squamate genus *Phacellaria* which, at first sight, might well be mistaken for an *Arceuthobium* of VISCACEAE. The extraordinary MISODENDRACEAE of southern S America must similarly be thought of as being mistletoe-like in habit. Thus the conception of "mistletoe" within SANTALALES has assumed a more ecological rather than precise taxonomic cachet, signifying plants which parasitized branches of woody hosts. Ironically, even this circumscription cannot adequately cover LORANTHACEAE s.s., as its most primitive trio of genera (*Atkinsonia*, *Nuytsia*, and *Gaiadendron*) are terrestrial root-parasites, only the latter having the ability to "ascend" a host tree, where it parasitizes fellow epiphytes, probably exclusively (Kuijt 1963).

[1] Department of Biology, University of Victoria, Victoria, B.C. V8W 3N5, Canada.

All illustrations, except Fig. 11, 12 and 21, are made by the author.

Acknowledgement. Financial support from the Natural Sciences and Engineering Council of Canada is gratefully acknowledged.

The above resumé does not exhaust contemporary controversy. A distinctive group of 3 genera (*Antidaphne*, *Eubrachion*, and *Lepidoceras*) which has traditionally been placed in VISCACEAE (or LORANTHACEAE subfam. VISCOIDEAE) were treated as a separate family, EREMOLEPIDACEAE (Kuijt, 1988). There is general agreement today that this is a natural assemblage, but molecular information has been promoted to indicate that EREMOLEPIDACEAE – and even VISCACEAE! – are better placed in SANTALACEAE, these ideas already having been incorporated in the third edition of Maas & Westra's "Neotropical plant families". Since such an alignment to me appears to be neither justified nor useful, the treatment which follows continues to recognize three families of mistletoes in the Guianas.

LITERATURE

Barlow, B.A. 1964. Classification of the Loranthaceae and Viscaceae. Proc. Linn. Soc. New South Wales ser. 2. 89: 268-272.

Barlow, B.A. & D. Wiens. 1971. The cytogeography of the Loranthaceous mistletoes. Taxon 20: 291-312.

Kuijt, J. 1963. On the ecology and parasitism of the Costa Rican tree mistletoe, Gaiadendron punctatum (Ruiz & Pavon) G. Don. Canad. J. Bot. 41: 929-938.

Kuijt, J. 1968. Mutual affinities of Santalalean families. Brittonia 20: 136-147.

Kuijt, J. 1969. The biology of parasitic flowering plants.

Kuijt, J. 1988. Monograph of the Eremolepidaceae. Syst. Bot. Monogr. 18: 1-60.

Maas, P.J.M. & L.Y.T. Westra. 2005. Neotropical plant families, ed. 3.

Wiens, D. & B.A. Barlow. 1971. The cytogeography and relationships of the viscaceous and eremolepidaceous mistletoes. Taxon 20: 313-332.

KEY TO THE FAMILIES OF GUIANAN MISTLETOES

1 Leafy plants; flowers often large and colorful, then (3-)5- or 6-merous, bi- or unisexual, ovary topped by a calyculus; inflorescence spicate, racemose, or umbellate, flowers often pedicellate and/or in triads/dyads and provided with bracteoles . . . . 105b. LORANTHACEAE

Leafy or squamate plants; flowers small and greenish, unisexual, 2- to 4-merous, ovary lacking a calyculus; inflorescences spicate, flowers sessile, ebracteolate, not in triads but often spread along, and sunken in internodes of spike 2

2 Leafy or squamate plants; flowers sunken into inflorescence axis, often in longitudinal series along inflorescence internodes; inflorescences usually axillary; leaves decussate . . . . 106. VISCACEAE

Squamate plants; flowers not sunken into inflorescence axis, solitary in axils of scale leaves; inflorescences at tips of branches; leaves alternate . . . . 105a. EREMOLEPIDACEAE

105a. EREMOLEPIDACEAE

by

JOB KUIJT

Semiparasitic mistletoes on branches of shrubs and trees, some with epicortical roots at base of plant, primary haustorium often becoming large and saddle-like. Leaves alternate, simple, short-petiolate to nearly sessile, expanded or scale-like and peltate. Dioecious or monoecious. Inflorescences spikes or catkin-like racemes; flowers pedicellate or sessile, apetalous in male *Antidaphne* only, where with central glandular disk and 2-4 stamens. Female and bisexual inflorescences spicate; flowers with 2 or 3 minute, caducous perianth segments. Fruit a small, 1-seeded berry; seed with abundant, whitish endosperm.

Distribution: A strictly American family ranging from Oaxaca in Mexico to northern Argentina and southern Chile, as well as the Greater Antilles, with 3 small genera totalling perhaps 12 species; in the Guianas 1 species.

Note: In addition to the genus here included, it is conceivable that *Eubrachion gracile* Kuijt occurs in the mid-western portion of Guyana, at elevations between 1400 and 2300 m. It is a slender, erect, squamate, monoecious mistletoe with small, cone-like inflorescences. It appears to be a Venezuelan endemic known mostly from the Auyan-tepui area about 100 km W of the Guyana border.

LITERATURE

Kuijt, J. 1988. Monograph of the Eremolepidaceae. Syst. Bot. Monogr. 18:1-60.

Kuijt, J. 1998. Eremolepidaceae. In J.A. Steyermark *et al.*, Flora of the Venezuelan Guayana 4: 731-734.

1. **ANTIDAPHNE** Poepp. & Endl., Nov. Gen. Sp. 2: 70. 1838.
   Type: A. viscoidea Poepp. & Endl.

Glabrous, foliaceous plants. Stems percurrent; basal epicortical roots present, with both primary and secondary haustoria. Leaves alternate, simple, variable in shape, shiny and somewhat leathery when fresh. Dioecious.

Distribution: 7 or 8 species, ranging from Mexico (Oaxaca) to southern Chile, 1 species being endemic to Cuba and Hispaniola; 1 species in the Guianas.

1. **Antidaphne viscoidea** Poepp. & Endl., Nov. Gen. Sp. 2: 70, t. 199. 1838. Type: Venezuela, Distr. Federal, Depto. Libertador, Steyermark 127742 (neotype MO, designated by Kuijt 1988: 33). – Fig. 1

Stachyphyllum fendleri Tiegh., Bull. Soc. Bot. France 42: 565. 1896. – *Antidaphne fendleri* (Tiegh.) Engl. in Engl. & Prantl, Nat. Pflanzenfam. Nachtr. II-IV, 1: 138. 1897. Type: Venezuela, Mérida, Fendler 1125 (holotype P).

Plants to ca. 1.5 m in diam. Stems slightly ridged, becoming terete; epicortical roots present at least when young, brownish, bearing conspicuous secondary haustoria; primary haustorium becoming large, often fusing with adjacent secondary haustoria in age. Leaves sessile or with short, indistinct petiole; blade mostly obovate with rounded apex. Male inflorescence a racemose catkin with caducous scales, terminal leaves absent. Female inflorescence spicate, flowers with semi-inferior ovary, tip of inflorescence with minute leaves which may later expand and become foliaceous; style short, stigma capitate to cristate; female inflorescence usually elongating after anthesis. Fruit grey-green to whitish or yellowish, ca. 0.4 cm long.

Distribution: Mexico (Oaxaca) to the Andes of northern Bolivia, northern and eastern Venezuela. In the Guianas known from a single collection from the Cuyuni-Mazaruni Region of Guyana (Pipoly *et al.* 10686 (LEA, NY)). The plant is very inconspicuous, however, and should be looked for elsewhere (GU: 1).

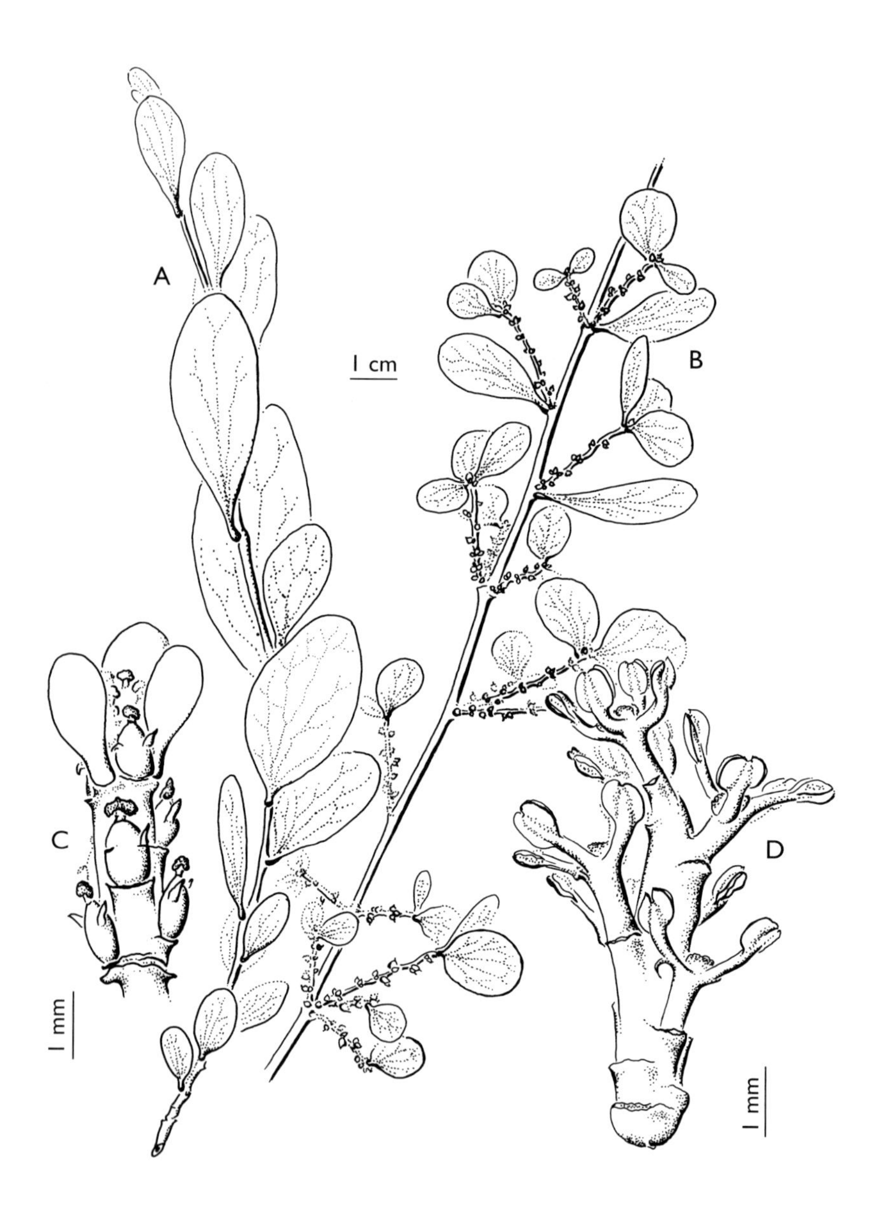

Fig. 1. *Antidaphne viscoidea* Poepp. & Endl.: A, young shoot; B, flowering shoot, female; C & D, female and male inflorescence, respectively (modified from Kuijt, Syst. Bot. Monogr. 18, 1988).

105b. LORANTHACEAE

by

JOB KUIJT

Leafy shrubs parasitic on other woody plants. Epicortical roots in a number of species, mostly at base, sometimes also from stems. Leaves alternate, decussate or whorled, Inflorescences extremely variable; flowers mostly in monads, dyads, or triads with fused bracts and/or bracteoles. Flowers mostly perfect, dioecious in some species, then flowers with aborted organs of opposite sex; calyx reduced to a very inconspicuous calyculus; petals conspicuous, (3-)4-7, fused or not, often brightly colored when flower perfect; stamens epipetalous, anthers dorsi- or basifixed, mostly dimorphic; style in many genera slender and long. Fruit a 1-seeded berry of diverse colors and size; seed endospermous in all but *Psittacanthus*. (x = 8, 9, 10, 11, 12, 24).

Distribution: About 700 species distributed over some 60 genera, largely in the tropics and subtropics of all continents; 15 genera in the New World; 37 species in 8 genera in the Guianas.

Notes: In the majority of New World LORANTHACEAE, stamens are of 2 sets, inserted at different heights on the subtending petals, reaching different heights, and often of somewhat different shapes. This is true even for such small-flowered genera as *Oryctanthus*, *Oryctina*, and *Phthirusa*. Additionally, some species exhibit considerable sexual dimorphism, non-functional stamens being present in female flowers and a non-functional style in the male.
Loranthus patrisii DC. (Prodr. 4: 288. 1830) does not belong to the Loranthaceae. Its type (French Guiana, Patris s.n., G-DC) is a species of *Combretum* (Combretaceae), probably *C. laxum* Jacq.

LITERATURE

Barlow, B.A. 1964. Classification of the Loranthaceae and Viscaceae. Proc. Linn. Soc. New South Wales ser. 2. 89: 268-272.
Barlow, B.A. & D. Wiens. 1971. The cytogeography of the Loranthaceous mistletoes. Taxon 20: 291-312.
Barlow, B.A. & D. Wiens. 1973. The classification of the generic segregates of Phrygilanthus (= Notanthera) of the Loranthaceae. Brittonia 25: 26-39.

Eichler, A.W. 1868. Loranthaceae. In C.F.P. von Martius, Flora Brasiliensis 5(2): 1-135.
Engler, H.G.A. & K. Krause. 1935. Loranthaceae. In H.G.A. Engler & K.A.E. Prantl, Die natürlichen Pflanzenfamilien, ed. 2. 16b: 98-203.
Krause, K. 1932. Loranthaceae. In A.A. Pulle, Flora of Suriname 1(1): 4-24.
Kuijt, J. 1964. A revision of the Loranthaceae of Costa Rica. Bot. Jahrb. Syst. 83: 250-326.
Kuijt, J. 1968. Mutual affinities of Santalalean families. Brittonia 20: 136-147.
Kuijt, J. 1978. Commentary on the mistletoes of Panama. Ann. Missouri Bot. Gard. 65: 736-763.
Kuijt, J. 1980. Miscellaneous mistletoe notes, 1-9. Brittonia 32: 518-529.
Kuijt, J. 1981. Inflorescence morphology of Loranthaceae - an evolutionary synthesis. Blumea 27: 1-73.
Kuijt, J. 1982. Seedling morphology and its systematic significance in Loranthaceae of the New World, with supplementary comments on Eremolepidaceae. Bot. Jahrb. Syst. 103: 305-342.
Kuijt, J. 1986. Loranthaceae. In G. Harling & B. Sparre, Flora of Ecuador 24: 113-198.
Kuijt, J. 1987. Miscellaneous mistletoe notes, 10-19. Brittonia 39: 447-459.
Kuijt, J. 1990. New species and combinations in neotropical mistletoes (Loranthaceae and Viscaceae). Proc. Kon. Ned. Akad. Wetensch., Biol. Chem. Geol. Phys. Med. Sci. 93: 113-162.
Kuijt, J. 1991. Inflorescence structure and generic placement of some small-flowered species of Phthirusa (Loranthaceae). Syst. Bot. 16: 283-291.
Kuijt, J. 1994. Typification of the names of New World mistletoe taxa (Loranthaceae and Viscaceae) described by Martius and Eichler. Taxon 43: 187-199.
Kuijt, J. 2001. Loranthaceae. In J.A. Steyermark *et al.*, Flora of the Venezuelan Guayana 6: 37-59.
Kuijt, J. 2003a. Miscellaneous mistletoe notes, 37-47. Novon 13: 72-88.
Kuijt, J. 2003b. Two new South American species of Struthanthus (Loranthaceae) posing a challenge to circumscription of neotropical genera. Bot. J. Linn. Soc. 142: 469-474.
Kuijt, J. & D. Lye. 2005. A preliminary survey of foliar sclerenchyma in neotropical Loranthaceae. Blumea 50: 323-355.
Lemée, A.M.V. 1955. Loranthacées. Flore de la Guyane Française 1: 535-544.
Lindeman, J.C. & A.R.A. Görts-van Rijn. 1968. Loranthaceae. In A.A. Pulle & J. Lanjouw, Flora of Suriname, Additions and corrections 1(2): 295-300.

Rizzini, C.T. 1956. Pars specialis prodromi monographiae Loranthacearum Brasiliae terrarumque finitimarum. Rodriguésia 18-19(30-31): 87-264.
Rizzini, C.T. 1978. Los generos venezolanos y brasileros de las Lorantáceas. Rodriguésia 30(46): 27-31.

KEY TO THE GENERA

1 Flowers 4-merous (rarely 3- or 5-merous), petals < 0.5 cm long 2
Flowers 6(-7)-merous, petals 0.1 cm to several cm long 3

2 Inflorescence determinate, single terminal flower following at least one pair of monads . . . . *1. Cladocolea*
Inflorescence indeterminate (i.e., lacking a terminal flower), consisting of lateral, bracteolate triads only . . . . *5. Phthirusa*

3 Flowers usually bright yellow (not red), each subtended by a separate green, foliar bract(eole); known from Mt. Roraima summit only . . . . *2. Gaiadendron*
Flowers not bright yellow, and not separately subtended by foliar bract(eole)s; diverse localities 4

4 Inflorescence with lateral monads only, including sessile monads 5
Inflorescence with lateral triads or dyads only 6

5 Flowers mostly partially immersed in inflorescence axis, each flanked by 2 strap-like, nearly obscured, bracteoles; pollen with 3 prominent excavations on each face . . . . *3. Oryctanthus*
Flowers not immersed in inflorescence axis; bracteoles not strap-like, ca. boat-shaped, largely exposed; pollen lacking prominent excavations . . . . *4. Oryctina*

6 Petals > 4 cm long, usually reddish or red and yellow; seed lacking endosperm . . . . *6. Psittacanthus*
Petals < 2 cm long, white, cream-colored or red; seed with endosperm . . .7

7 Flowers unisexual, plants dioecious 8
Flowers bisexual 9

8 Anthers basifixed, longer filaments laterally excavated to accommodate lower anthers . . . . *5. Phthirusa*
Anthers dorsifixed, versatile, filaments slender, not laterally excavated . . . . *7. Struthanthus*

9 Triads and flowers stalked; anthers dorsifixed, versatile . . . *8. Tripodanthus*
Triads and flowers sessile or essentially so; anthers basifixed, longer filaments laterally excavated to accommodate lower anthers . . . . *5. Phthirusa*

1. **CLADOCOLEA** Tiegh., Bull. Soc. Bot. France 42: 166. 1895.
Type: C. andrieuxii Tiegh.

Small plants, glabrous to furfuraceous when young; at least sometimes with a few slender epicortical roots from base of plant. Leaves irregularly alternate-decussate. Inflorescences determinate or indeterminate, extremely small, often (but not always) with bracteolate triads below and ebracteolate monads above, all but terminal flowers in paired positions. Flowers bisexual or plants dioecious; 3-5-merous; stamens/anthers fused with subtending petals, dimorphic, with 2 or 4 pollen sacs.

Distribution: About 40 species, mostly Central Mexico, S America to northern Peru and Bolivia; in the Guianas 3 species.

KEY TO THE SPECIES

1 Flowers 5-merous, leaves shiny, with evident venation . . . . *2. C. nitida*
Flowers 3-4-merous, leaves dull, with inconspicuous to obscure venation 2

2 Anther a thin structure with 4 minute pollen sacs at its corners, completely fused with petal . . . . *3. C. sandwithii*
Anther 2-lobed, incompletely fused with petal . . . . *1. C. micrantha*

1. **Cladocolea micrantha** (Eichler) Kuijt, Syst. Bot. 16: 288. 1991. – *Phthirusa micrantha* Eichler in Mart., Fl. Bras. 5(2): 65, t. 19, f. 9. 1868. – *Passovia micrantha* (Eichler) Tiegh., Bull. Soc. Bot. France 42: 172. 1895, as 'Passowia'. – *Struthanthus micranthus* (Eichler) Baehni & J.F. Macbr., Candollea 7: 290. 1937. Type: Brazil, Amazonas, Spruce 1782 (holotype M). – Fig. 2

Phthirusa micrantha Eichler var. *bolivariensis* Rizzini, Rodriguésia 28(41): 12. 1976. Type: Venezuela, Bolívar, Steyermark 86879 (holotype RB, not seen).
Phthirusa bernardiana Rizzini, Rodriguésia 28(41): 12. 1976. Type: Venezuela, Bolívar, Bernardi 2840 (holotype RB, not seen).
Cladocolea elliptica Kuijt, Novon 2: 354. 1992. Type: Suriname, Irwin *et al.* 57535 (holotype P, isotypes BBS, NY), syn. nov.

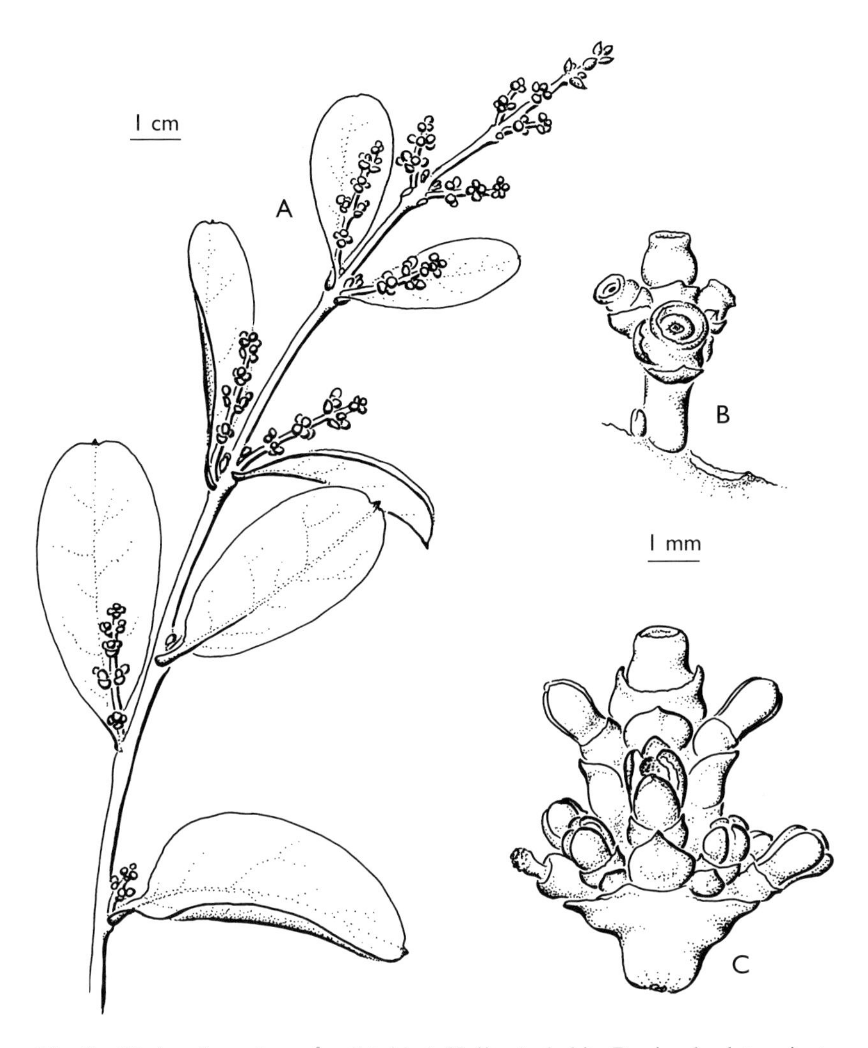

Fig. 2. *Cladocolea micrantha* (Eichler) Kuijt: A, habit; B, simple determinate inflorescence with monads only; C, determinate inflorescence with basal triads (A, based on Clarke *et al*. 7329; B-C, reproduced from Kuijt, Syst. Bot. 16, 1991).

Plants with several slender epicortical roots from base; young stem and inflorescence brown, furfuraceous; internodes terete, to 9 cm long. Leaves irregularly paired; petiole 0.2-0.5 cm long; blade extremely variable, lanceolate, elliptic, or ovate, to 8 x 3.5 cm, apex rounded to deeply emarginate, usually minutely apiculate, base obtuse to cuneate; venation ranging from pinnate with a distinct midvein reaching into apex to 3-veined in more emarginate leaves. Dioecious, male plants apparently infrequent. Inflorescence determinate, 0.5-3 cm long, apical flower followed by one or more pairs of bracteolate or ebracteolate, sessile monads, below which sometimes one or more pairs of sessile triads; peduncle 0.1 cm, often elongating in fruit; secondary inflorescences frequent in a superposed position; vigorous plants also with whip-like terminal, compound inflorescence to 15(-20) cm long, where lateral inflorescences subtended by caducous, fleshy, awl-shaped scale leaves of 0.2-0.3 cm long. Male and female flowers similar, 3- or 4-merous, < 0.2 cm long; petals and stamens slightly dimorphic, organs of opposite sex small, anthers sessile, 2-loculate; style straight, stigma poorly differentiated. Fruit ellipsoid or ovoid, 0.5 x 0.3 cm, light orange to red with dark tip.

Distribution: Venezuela (Bolívar) through the Guianas and into northern Brazil; 30+ collections studied, 16 from the Guianas (GU: 14; SU: 1; FG: 1).

Selected specimens: Guyana: Rupununi Distr., Shea Rock, Jansen-Jacobs *et al.* 3697 (LEA, U); Upper Takutu-Upper Essequibo Region, Acarai Mts., Clarke *et al.* 7329 (LEA, US). French Guiana: Piste de Saint-Elie, Interfluve Sinnamary-Counamama, Prévost *et al.* 3408 (CAY, LEA, NY, K, MO).

Notes: Plants with larger, strictly pinnately veined leaves are very similar to *Phthirusa guyanensis*, which has indeterminate inflorescences occasionally elongating to 6 cm, the bracts and bracteoles forming a more united, densely furfuraceous cupule, and on which members of triad pairs tend to become widely separated from each other.
Cladocolea elliptica is believed to fall within the considerable range of variability of *C. micrantha*; its stated indeterminate inflorescences need to be confirmed. There is a form of the species with extremely narrow, lanceolate leaves with acute leaf base.

2. **Cladocolea nitida** Kuijt, Novon 13: 74. 2003. Type: Guyana, Potaro-Siparuni Region, Pakaraima Mts., Henkel *et al.* 4400 (holotype LEA, isotype US).

Sparsely branched plants, glabrous; internodes quadrangular when young. Leaves paired; slender petiole ca. 1 cm long; blade obovate, to 4 x 2 cm, apex rounded to notched or nearly mucronulate, base acute, upper surface shiny when dry, lower surface dull; pinnate venation clearly marked on both sides. Dioecious, only female plants known.

Female inflorescence, including flowers, 0.5 cm long, spicate, with 5 ebracteolate flowers, at least lowest ones subtended by a caducous bract 0.5 cm long, the 5 flowers in two pairs plus one terminal one. Flowers 0.45 cm long, calyculus entire, flaring; petals 5, each with a minute, elongate, staminodial cushion; ovary 0.1 cm long, style nearly as long as petals, surrounded at base by nectary disk, stigma capitate, very prominent.

Distribution: Known from the type only (GU: 1).

3. **Cladocolea sandwithii** (Maguire) Kuijt, Syst. Bot. 16: 289. 1991. – *Phthirusa sandwithii* Maguire in Maguire *et al.*, Bull. Torrey Bot. Club 75: 302. 1948. Type: Guyana, Kaieteur Savanna, Sandwith 1404 (holotype NY, isotypes K, U).

Internodes terete, furfuraceous, rather slender, 2-4 cm long. Leaves irregularly paired; petiole rather slender, 0.5 cm long; blade somewhat coriaceous, obovate, to 5 x 1.5 cm, apex emarginate and mucronate to rounded, base acute, upper surface of blade showing several lateral, basal veins, lower surface with furfuraceous midrib to apex; margins and petioles furfuraceous. Inflorescence ca. 0.3 cm long, sessile, 1-3 per axil, usually with 4 triads and one terminal flower, upper 2 triads sometimes replaced with monads; bracts somewhat furfuraceous. Dioecious, only male plants seen. Flowers (male) 4-merous, ca. 0.2 cm long; calyculus irregularly lacerate; anthers fused with petals, each reduced to 4 minute pollen sacs, upper 2 placed well above others; style conspicuously clavate, slightly more than 0.1 cm long.

Distribution: Known from the type only (GU: 1).

Note: A species with very close affinities to *C. micrantha*, but with very different anther structure, the anther no more than a thin cushion with 4 extremely small pollen sacs at the corners.

2. **GAIADENDRON** G. Don, Gen. Hist. 3: 431. 1834.
Type: G. punctatum (Ruiz & Pav.) G. Don (Loranthus punctatus Ruiz & Pav.)

Trees or shrubs of usually terrestrial habitat, parasitic on roots of other vascular plants. Leaves dark green, paired, broadly lanceolate to ovate. Inflorescences a raceme of stalked triads, median flowers of triads sessile, lateral ones short-stalked, each subtended by a green, foliaceous bract/bracteole. Flowers bisexual, 6- or 7-merous; anthers versatile and dorsifixed. Fruit a rather large, pulpy berry; seed not viscid, with grooved, white endosperm and dicotylous, small embryo. (x = 12, 24).

Distribution: See under the species.

Note: Presently regarded as a monotypic genus.

1. **Gaiadendron punctatum** (Ruiz & Pav.) G. Don, Gen. Syst. 3: 431. 1834. – *Loranthus punctatus* Ruiz & Pav., Fl. Peruv. Chil. 3: 47, t. 276, f. a. 1802. – *Phrygilanthus punctatus* (Ruiz & Pav.) Eichler in Mart., Fl. Bras. 5(2): 48. 1868. Type: Peru, Pavon s.n. (holotype MA, not seen, GH photo 29459, isotypes BM, G, MO, P).
– Fig. 3

Shrub or small tree, mostly terrestrial, sometimes epiphytic on trees while parasitizing other epiphytes, to 10 m high or more. Branches more or less terete. Leaves in pairs, shiny, leathery; petiole about 1 cm long, distinct; blade broadly lanceolate to ovate, to 8 x 4 cm, apex slightly attenuate, margin somewhat revolute below. Inflorescence solitary in leaf axils and terminating branch, a raceme of stalked triads (sometimes also 2-4 monads and single, terminal flower); each triad with sessile median flower, and green, foliaceous bracts and bracteoles. Flowers commonly golden yellow (rarely white, especially in central Colombia); petals mostly 0.8-2.5 cm long; stamens epipetalous, of two or more lengths, anthers versatile, dorsifixed; style straight, stigma scarcely differentiated. Fruit a dull orange berry lacking viscin, spherical, about 1 cm in diam.; mature seed with white endosperm which has radiating flanges, embryo dicotylous, narrow, 0.2 cm long, lacking a haustorial disk; germination epigeous.

Distribution: Nicaragua through Central America to Bolivia along the Andes (the Guyana collections being the easternmost ones known); several hundred collections studied, 4 from Guyana (GU: 4).

Selected specimens: Guyana: Upper Potaro R. Region, summit of Mt. Wokomung, Boom *et al.* 9081 (NY); Cuyuni-Mazaruni Region, 2-5 km NW of "prow" of Mt. Roraima, Hahn *et al.* 5451 (LEA, US).

Note: Characteristic as a small terrestrial tree at the lower edges of páramo vegetation in most of its range but, in its "epiphytic" form (when parasitic on other epiphytes), may also occur at somewhat lower levels in context of moist forests.

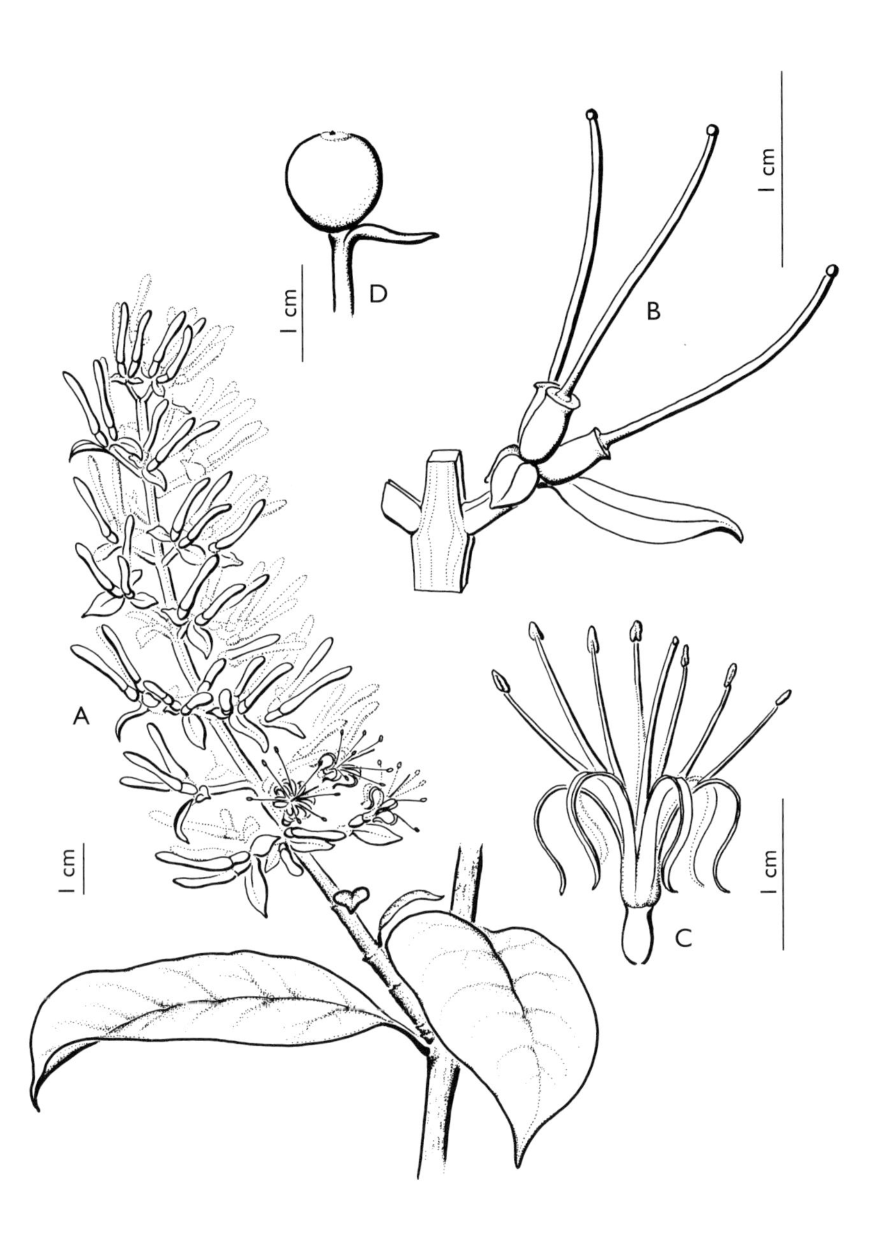

Fig. 3. *Gaiadendron punctatum* (Ruiz & Pavon) G. Don: A, habit; B, inflorescence triad shortly after flowering; C, flower; D, fruit (reproduced from Kuijt, Fl. of Ecuador 24, 1986).

3. **ORYCTANTHUS** (Griseb.) Eichler in Mart., Fl. Bras. 5(2): 87. 1868. – *Loranthus* sect. *Oryctanthus* Griseb., Fl. Brit. W. I. 313. 1860. Type: O. occidentalis (L.) Eichler (Loranthus occidentalis L.)

Leafy parasitic shrubs, glabrous but young twigs often strongly furfuraceous; epicortical roots from base of plant but not from stems, sometimes lacking at least at maturity. Leaves paired, often leathery; venation palmate or pinnate, containing characteristic stellate fiber bundles along smaller veins. Inflorescences solitary in leaf axil or sometimes clustered, sometimes arranged in a terminal, compound, squamate inflorescence, individually a spike, flowers sessile in cavities of often swollen axis, each flower subtended by 2 minute, acute, strap-shaped bracteoles. Flowers bisexual, small, 6-merous (in the Guianas species), yellow to dark red; petals and stamens dimorphic; pollen with 3 circular depressions on each face; style straight, with scarcely differentiated stigma. Fruit green, yellow-green, red to purple or black; endosperm copious, white or yellowish; mature seed with fleshy, dicotylous embryo, cotyledons massive, haustorial disk well differentiated.

Distribution: Southern Mexico (Tabasco & Chiapas), through Central America to north-eastern Bolivia and the northern half of Brazil, with a population of *O. occidentalis* on Jamaica; taxonomically distinctive genus with 11 often inconspicuous species; 3 in the Guianas.

Notes: In addition to the 3 species treated below, it is possible that *O. spicatus* (Jacq.) Eichler also occurs in the Guianas. It is very close to *O. florulentus* (in fact, it may intergrade with it W of the Guianas) but is more slender and usually smaller, its spikes having rather long and slender peduncles, and its fruits somewhat smaller and more rounded at the tip.
An additional species, *O. cordifolius* (C. Presl) Urb., was previously reported for Guyana (Kuijt 1976), but this appears to be an error.

LITERATURE

Kuijt, J. 1961. Notes on the anatomy of the genus Oryctanthus (Loranthaceae). Canad. J. Bot. 39: 1809-1816.
Kuijt, J. 1976. Revision of the genus Oryctanthus (Loranthaceae). Bot. Jahrb. Syst. 95: 478-534.
Kuijt, J. 1992. Nomenclatural changes, new species, and a revised key for the genus Oryctanthus (Loranthaceae). Bot. Jahrb. Syst. 114: 173-183.

KEY TO THE SPECIES

1 Flowers, buds, and fruits perpendicular to the spike or nearly so; inflorescences axillary only . . . . *3. O. occidentalis*
Flowers, buds, and fruits clearly inclined forward on the spike; inflorescences axillary or in open, terminal, compound groups, or both 2

2 Young stems angular, with brown furfuraceous stripes . . . . *2. O. florulentus*
Young stems terete, lacking furfuraceous stripes; inflorescence axillary or in terminal compound inflorescences, or both . . . . *1. O. alveolatus*

1. **Oryctanthus alveolatus** (Kunth) Kuijt, Bot. Jahrb. Syst. 95: 504. 1976. – *Loranthus alveolatus* Kunth in Humb., Bonpl. & Kunth, Nov. Gen. Sp. ed. qu. 3: 444. 1820. Type: Colombia, Humboldt & Bonpland s.n. (holotype P-Bonpl.). – Fig. 4

Oryctanthus botryostachys Eichler in Mart., Fl. Bras. 5(2): 89. 1868. Type: Brazil, Pará, Spruce 735 (lectotype M, designated by Kuijt 1976: 510).
Oryctanthus alveolatus (Kunth) Kuijt var. *kuijtii* Rizzini in Luces & Steyerm., Fl. Venez. 4(2): 136. 1982. Type: Venezuela, Bolívar, Steyermark 90482 (holotype VEN).

Rather large plant; stems stout, terete. Petiole stout, several mm long, or leaves clasping; blade varying from orbicular to cordate, broadly ovate or rather narrowly elliptical, to 7 x 6 (14 x 11 cm), but commonly smaller; venation clearly palmate at least in broad-leaved forms, midvein not reaching apex. Inflorescence to 5 cm long, when axillary 1-several per axil, often also in terminal, panicle-like compound groups where subtended by scale-leaves; peduncle stout, to 1 cm long but sometimes nearly absent; spikes normally swollen, flowers 30-60 per spike, scale leaves often glabrous. Flowers strongly inclined forward, ca. 0.3 cm long, slightly less than half of which is ovary; petals rather narrow, 0.2 cm long, strongly dimorphic; anthers inserted at or above middle of petals, 2 inner pollen sacs much smaller than outer ones, shortest anther with attenuate connective; style reaching to upper anthers, stigma broadly capitate. Fruit greenish, apparently becoming reddish purple, sometimes with yellow base, ovoid, inclined forward at about 45°.

Distribution: Ranges from at least Costa Rica to Peru, Bolivia, and Brazil; common locally throughout the Guianas, but never very conspicuous; ca. 75 collections studied, ca. 20 from the Guianas (GU: 15+; SU: 5; FG: 5).

Fig. 4. *Oryctanthus alveolatus* (Kunth) Kuijt: A, habit; B, inflorescence (reproduced from Kuijt, Fl. of Ecuador 24, 1986).

Selected specimens: Guyana: Demerara Compartment, CD 920 road, Ek 748 (U); Cuyuni-Mazaruni Region, Cuyuni R., between Aurora and Takar-opati Island, Gillespie *et al.* 2335 (LEA, US). Suriname: Saramacca, Coppename R. near Raleighfalls Island, Lanjouw 967 (U); Tramway Falls, Corantijn R., Rombouts 152 (U). French Guiana: St. George d'Oyapoque, Lemée s.n. (F); St. Laurent-du-Maroui, Wurdack 4045 (NY).

Note: *O. alveolatus* is a variable species insufficiently known for infrageneric categories to be convincing. The plant is an important, but frequently ignored, tree pathogen in many areas of its geographic range.

2. **Oryctanthus florulentus** (Rich.) Tiegh., Bull. Mus. Hist. Nat. (Paris) 2: 339. 1896. – *Loranthus florulentus* Rich., Actes Soc. Hist. Nat. Paris 1: 107. 1792. Type: French Guiana, Cayenne, Leblond 222 (holotype G) (not Richard s.n. as indicated in Kuijt 1976).

Viscum pennivenium DC., Prodr. 4: 282. 1830. – *Phoradendron pennivenium* (DC.) Eichler in Mart., Fl. Bras. 5(2): 128. 1868. Type: French Guiana, Perrottet 227 (G).
Loranthus ruficaulis Poepp. & Endl., Nov. Gen. Sp. 2: 61, t. 185. 1838. – *Oryctanthus ruficaulis* (Poepp. & Endl.) Eichler in Mart., Fl. Bras. 5(2): 90. 1868. Type: Brazil, Amazonas, Poeppig s.n. (holotype BM).
Loranthus vestitus Miq., Linnaea 18: 62. 1844. Type: Suriname, Focke 607 (holotype U).
Loranthus surinamensis Miq., Linnaea 18: 63. 1844. Type: Suriname, Focke 406 (holotype U).
Loranthus chloranthus Miq., Stirpes Surinam. Sel. 205. 1851. – *Oryctanthus chloranthus* (Miq.) Eichler in Mart., Fl. Bras. 5(2): 89. 1868. Type: Suriname, Focke s.n. (holotype U, isotypes GH, K).
Oryctanthus ruficaulis (Poepp. & Endl.) Eichler var. *latifolius* Eichler in Mart., Fl. Bras. 5(2): 91. 1868, as 'latifolia'. Type: Brazil, Pará, Burchell 9211 (lectotype BR, designated by Kuijt 1994: 189).

Rather small-leaved plant, sparsely branched; stems sharply angular until older, when becoming more or less terete; internodes usually 3 cm long or less; youngest twigs with very conspicuous furfuraceous lines continuing along lower midrib of each leaf into apex or nearly so. Petiole 0.3 cm long; blade rather thin, broadly lanceolate, mostly about 4 x 2 cm, rarely to 7.5 x 3.5 cm, rounded at apex, acute at base; venation pinnate. Spikes axillary only, 1-3 per axil, 2-4 cm long; peduncle less than 0.2 cm, floriferous portion rather slender, about 24 flowers or more per spike; bracteoles extending above bract. Flowers greenish, orange, or light red, somewhat angular, ca. 0.2 cm long; anthers strongly dimorphic, 4-

loculate, lower anthers nearly touching nectary, upper anthers at about middle of petal, connectival horns of 2 anther series little different; stigma reaching above upper anthers. Fruit orange-red, rather bluntly elliptical, nearly perpendicular to axis.

Distribution: From Colombia and Amazonian lowlands of Ecuador and Peru through Brazil and the Guianas; mostly at low elevations; 100+ collections studied, ca. 67+ from the Guianas (GU: ca. 25; SU: 30+; FG: 12).

Selected specimens: Guyana: Port Kaituma, on Guattaria, Polak *et al.* 87 (LEA, U); Essequibo-W Demerara Region, Naamryck Canal, just W of Lookout Village, Pipoly 11277 (LEA, US). Suriname: Edge of savanna, a few km from Zanderij, on Myrtaceae, Kramer & Hekking 2602 (U); Tafelberg, Savanna II, Maguire 27202 (F, GH, K, MO, NY, U, US). French Guiana: Near Cayenne, Broadway 681 (GH, NY, US); Karouany, Sagot 297 (BM, BR, K, U).

Note: *O. florulentus* is a difficult species vis-à-vis *O. spicatus* (Jacq.) Eichler, which appears to intergrade with it; the peduncle of the latter tends to be much longer, and the plants smaller.

3. **Oryctanthus occidentalis** (L.) Eichler in Mart., Fl. Bras. 5(2): 89. 1868. – *Loranthus occidentalis* L., Amoen. Acad. 5: 396. 1760. Type: Jamaica, Browne s.n., (holotype BM-SL). – Fig. 5

   In the Guianas only: subsp. **continentalis** Kuijt, Bot. Jahrb. Syst. 114: 181. 1992. Type: Costa Rica, Puntarenas, Kuijt 2570 (holotype CR, isotype UBC).

Young stems and peduncles light cinnamon, furfuraceous, always terete. Leaves fairly thick but not leathery, variable in shape and venation, 5.5 x 3.5 cm (to 15 x 12 cm). Spikes axillary only, single or in clusters, 2-4(-12) cm long; peduncle 0.8-1.3 cm, floriferous part glabrous, large spikes bearing to ca.180 flowers. Flowers perpendicular to axis; anthers 4(2)-loculate. Fruit green or green with yellow or red, 0.3(0.2) x 0.2 cm, perpendicular to axis.

Distribution: Jamaica (whence the type), and Costa Rica to Peru and east into Suriname; all continental material belongs to subsp. *continentalis*; 50+ collections studied, of which ca. 10 from the Guianas (SU: ca. 8; FG: 2).

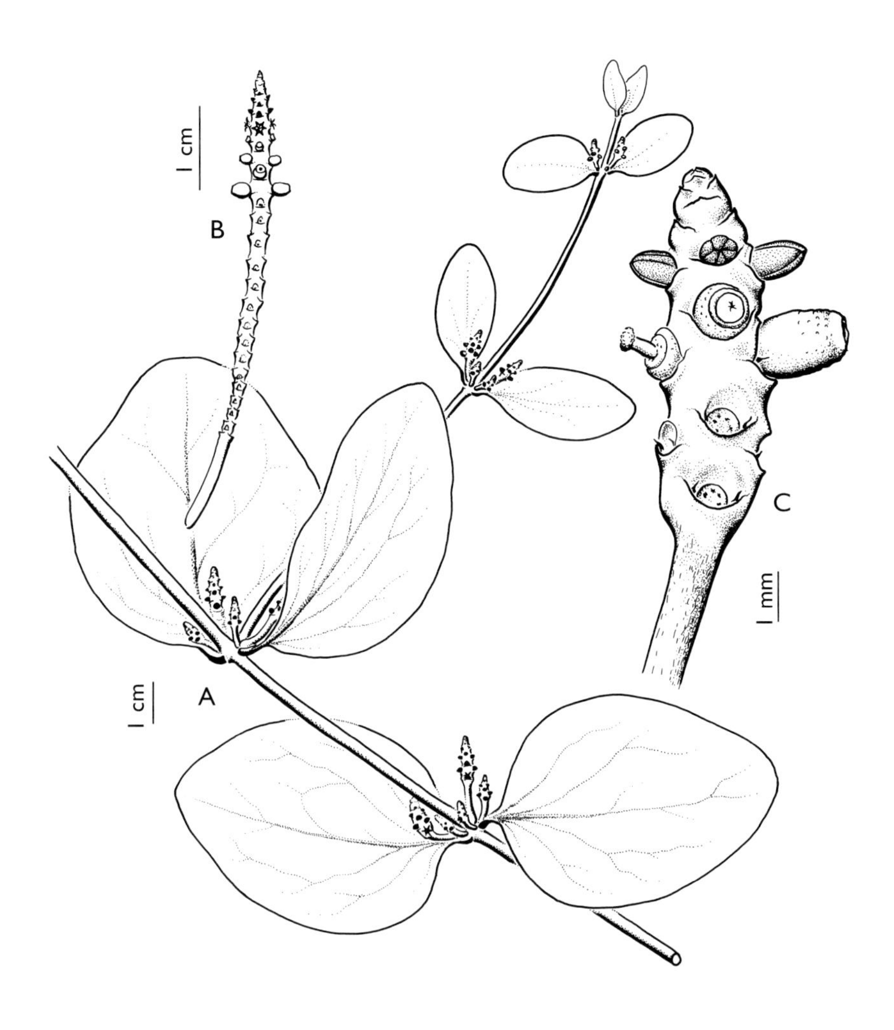

Fig. 5. *Oryctanthus occidentalis* (L.) Eichler: A, habit; B, older inflorescence; C, details of young inflorescence (reproduced from Kuijt, Fl. of Ecuador 24, 1986).

Selected specimens: Suriname: Suriname R., old sawmill of Suhoza, Lindeman 4630 (U); Jodensavanne-Mapane Kreek area, Lindeman 7270 (U). French Guiana: Vicinity of Cayenne, Broadway 846 (GH); without locality, Poiteau s.n. (K).

4. **ORYCTINA** Tiegh., Bull. Soc. Bot. France 42: 168. 1895.
Type: O. scabrida (Eichler) Tiegh. (Oryctanthus scabridus Eichler)

Leafy parasitic plants; at least some with basal epicortical roots. Leaves decussate, lacking stellate fiber bundles. Inflorescences indeterminate spikes bearing pairs of sessile flowers (monads); each flower with 2 (in Guianan species) prominent boat-shaped bracteoles extending well above subtending bract. Flowers bisexual or plants dioecious; 6-merous, sessile; petals and stamens dimorphic in each flower, anthers 2(4)-loculate, pollen triangular to slightly convex, psilate, syncolpate.

Distribution: A small genus presently of 9 species, Costa Rica (1 species in Osa Peninsula), Venezuela, the Guianas, Brazil (4 species in Goiás, Minas Gerais, and adjacent Bahia); in the Guianas 2 species.

Note: The similar genus *Oryctanthus* is consistently different in its unique pollen type, the stellate fiber bundles in its leaves, and the minute, flat, strap-shaped bracteoles.

LITERATURE

Kuijt, J. 1991. Inflorescence structure and generic placement of some small-flowered species of Phthirusa (Loranthaceae). Syst. Bot. 16: 283-291. 1991.

Kuijt, J. 2000. Two new species of Oryctina (Loranthaceae) with a revised key to the genus. Novon 10: 391-397.

Kuijt, J. 2003. A new species of Oryctina (Loranthaceae) from Guyana. Brittonia 55: 169-172.

Kuijt, J. (under review). First record of the genus Oryctina (Loranthaceae) in Mesoamerica: O. costaricensis, sp. nov. Novon.

KEY TO THE SPECIES

1 Leaves obovate to orbicular, lower midrib forming a distinct, dark line; upper part of style swollen in fusiform fashion at least in some flowers . . . . *1. O. atrolineata*

Leaves ovate, lower midrib lacking dark line; style not swollen . . . . *2. O. myrsinites*

1. **Oryctina atrolineata** Kuijt, Brittonia 55: 169. 2003. Type: Guyana, Demerara Region, Timehri, Cremers *et al.* 10912 (holotype CAY, isotype LEA). – Fig. 6

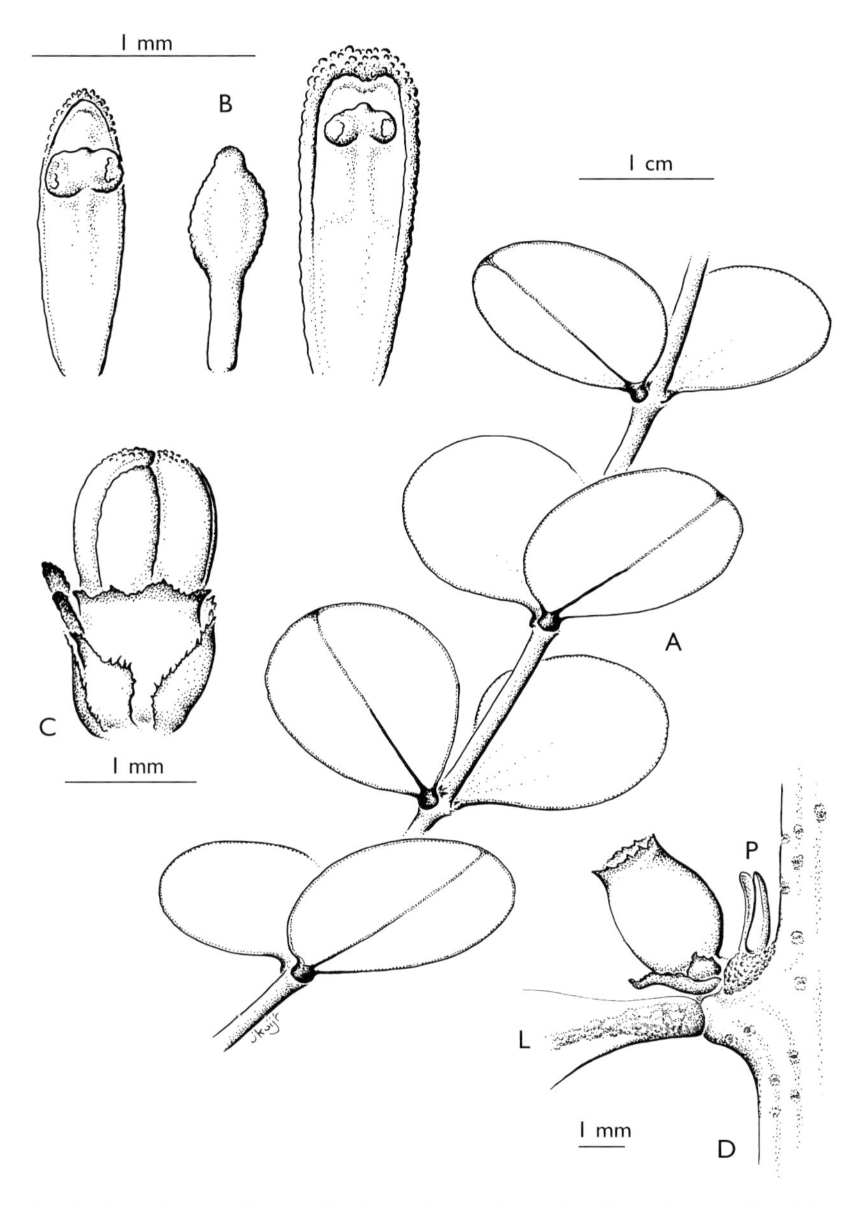

Fig. 6. *Oryctina atrolineata* Kuijt: A, leafy shoot; B, dimorphic petals with sessile 2-loculate anthers and (middle) style; C, mature flower bud, subtended by bract (1) and bracteoles (2); D, immature fruit, bract and 1 bracteole showing (L, subtending leaf; P, erect prophylls of the 1-flowered inflorescence) (reproduced from Kuijt, Brittonia 55, 2003).

Small plants, sparsely branched, with epicortical roots at base; internodes more or less terete, furfuraceous when young, glabrescent. Petiole 0.1 cm long; blade fleshy, broadly obovate to nearly orbicular, to 2 x 1.5 cm, apex rounded, often emarginate, base tapering abruptly to petiole; venation obscure except for lower midrib which, when dry, forms a straight, dark line continuous at apex with dark leaf margin. Flowers in small clusters in leaf axils, forming extremely short, sessile, 1-flowered spikes, each flower subtended by a scale-like bract and 2 minute, black-tipped bracteoles. Flowers unisexual?, 0.2 cm long; petals 6, 0.1 cm long, strongly dimorphic as are the minute, sessile, 2-loculate anthers; style swollen in fusiform fashion in its upper portion, stigma scarcely differentiated. Fruit barrel-shaped, 0.35 x 0.2 cm, waxy white when dry, with lacerate, brown calyculus.

Distribution: Known only from the type and 1 additional collection (GU: 2).

Specimens examined: Guyana: the type; E Demerara Region, Yarowkabra settlement and Forestry Commission Station, Pipoly *et al.* 7388 (LEA, NY).

Note: It is undoubtedly the very inconspicuous habit of the plant that has allowed it to escape notice in Guyana; it is likely to be more common.

2. **Oryctina myrsinites** (Eichler) Kuijt, Syst. Bot. 16: 290. 1991. – *Phthirusa myrsinites* Eichler in Mart., Fl. Bras. 5(2): 66. 1868. – *Passovia myrsinites* (Eichler) Tiegh., Bull. Soc. Bot. France 42: 172. 1895, as ‘Passowia’. Type: Brazil, Pará, Spruce 618 (lectotype P, designated by Kuijt 1991: 290).

   Phthirusa savannarum Maguire in Maguire *et al.*, Bull. Torrey Bot. Club 75: 301. 1948. – *Phthirusa myrsinites* Eichler var. *savannarum* (Maguire) Rizzini, Mem. New York Bot. Gard. 29: 25. 1978. Type: Suriname, Tafelberg, Savanna II, Maguire 24702 (holotype NY, isotypes U, US).

Sparsely branched plants with furfuraceous, slender internodes to 3 cm long, chocolate-brown; epicortical roots only from base of plant; older stems with longitudinal series of rough brown lenticels. Petiole 0.1-0.2 cm long, brown; blade obovate, mostly 2 x 1 cm, up to 4 x 2.5 cm, apex truncate to slightly emarginate, base acute, petiole, leaf margin and lower midrib brown, these contrasting sharply with grey-green blade; upper surface sometimes showing several faint palmate veins.

Inflorescence extremely short, spike-like, often several per axil, individual flowers sessile, subtending prophylls and bracts nearly 0.1 cm long, somewhat carinate, massive, glabrous, bracts caducous; bracteoles fimbriate along margin especially at tip. Flowers 0.15 cm long; petals and anthers dimorphic, anthers 2-loculate, alternatingly small and button-like and larger, cordate; ovary somewhat less than 0.05 cm long, calyculus brown, membranous, irregularly lacerate, style 0.05 cm, straight. Fruit ovoid, 0.3 x 0.2 cm, drying waxy white when mature, calyculus brown, sclerotic, torn, conspicuous on truncate apex.

Distribution: Venezuela (Bolívar), through the Guianas, adjacent Brazil; 5 specimens studied, 2 from the Guianas (SU: 1; FG: 1).

Specimens examined: Suriname: Tafelberg, Savanna II, Maguire 24702 (NY, U, US). French Guiana: Camp No.1, Ouman fou langa Soula, Bassin du Haut-Marouini, de Granville 9608 (CAY, P, US).

5. **PHTHIRUSA** Mart., Flora 13: 110. 1830.
Type: P. clandestina (Mart.) Mart. (Loranthus clandestinus Mart.)

Passovia H. Karst. in Klotzsch, Bot. Zeitung (Berlin) 10: 305. 1852.
Type: *P. suaveolens* H. Karst.
Furarium Rizzini, Arch. Jard. Bot. Rio de Janeiro 12: 118. 1953.
Type: *F. disjectifolium* Rizzini

Leafy parasites, glabrous or furfuraceous when young; usually epicortical roots from base of plant and, in some, also from branches. Leaves paired, petiolate. Inflorescences not subtended by bracts, basically an indeterminate raceme of sessile or pedunculate triads, each triad with at least median flower sessile, and with bracts and bracteoles which tend to be persistent; in some species, branched (compound) inflorescences in terminal and/or axillary positions. Flowers perfect or plants dioecious, 4- or mostly 6-merous, small, dark red to creamy white; when unisexual, aborted organs of opposite sex present; stamens dimorphic in most species, with basifixed anthers, filaments short and stout, longer ones often with lateral depressions accommodating anthers of shorter ones; style rather massive, with shape of anthers impressed upon it, and thus of uneven thickness. Fruit a berry of diverse colors; seed with copious white endosperm; embryo at maturity bright green, fleshy, dicotylous, with well differentiated haustorial disk. (x = 8, 16).

Distribution: Uncertain number of species (see note below); almost exclusively continental, S American genus (2 species on Jamaica), ranging from southern Mexico to Bolivia, only *P. pyrifolia* being present north of Costa Rica; there is a strong preference for low elevations in *Phthirusa*; in the Guianas 11 species.

Note: The closely related, strictly Caribbean genus *Dendropemon* (Blume) Rchb. has in the past frequently been synonymized under *Phthirusa*. The present genus is of uncertain size as it has not been analyzed properly since Eichler's treatment in the Flora Brasiliensis in 1868; it is probably not a natural assemblage. Its separation from *Struthanthus* is said to be difficult, but this is perhaps due to the lack of attention to the details of floral structure; no *Phthirusa* species has slender filaments and versatile anthers, as is the case in *Struthanthus*. Anther morphology in the two genera would seem to be very different, those of *Struthanthus* being dorsifixed and versatile, those of *Phthirusa* being basifixed (sometimes sessile), the connective of some species extending into a small horn and the filament massive and laterally excavated (longer stamen series). See also the comments under *Struthanthus*.

LITERATURE

Kuijt, J. 1991. Inflorescence structure and generic placement of some small-flowered species of Phthirusa (Loranthaceae). Syst. Bot. 16: 283-291.

Kuijt, J. & E.A. Kellogg. 1996. Miscellaneous mistletoe notes, 20-36 (see 34 & 35). Novon 6: 33-53.

Rizzini, C.T. 1952. Phthirusae Brasiliae terrarumque adiacentium. Dusenia 3: 451-462.

KEY TO THE SPECIES

1 Plants very finely pubescent; triads 2, peduncle extremely delicate; petals 4 . . . . *11. P. trichodes*

Plants glabrous throughout; triads mostly numerous, sessile or nearly so, or peduncles not delicate; petals 3-6 2

2 Plants very slender; leaves very narrowly lanceolate, mostly < 5 x 1 cm, venation obscure except for midrib; triads 2-4 pair per inflorescence . . . . *10. P. stenophylla*

Plants not unusually slender; leaves ovate or at least not narrowly lanceolate, mostly > 6 x 2 cm, venation pinnate, evident; triads mostly > 5 pairs per inflorescence 3

3 Inflorescence peduncles conspicuously flattened or winged . . . . *5. P. podoptera*

Inflorescence peduncles terete or only slightly angled 4

4 Inflorescence axis strongly brown-furfuraceous; leaf apex acute, base truncate or nearly so; dioecious . . . . *8. P. rufa*
Inflorescence axis glabrous or no more than very thinly furfuraceous; leaf apex acute to rounded, base obtuse to acute; dioecious or flowers bisexual 5

5 Leaf apex acute 6
Leaf apex obtuse to rounded 8

6 Flowers bisexual; epicortical roots from base only; stems compressed or even keeled when young, with furfuraceous lines . . . . *7. P. pyrifolia*
Plants dioecious; epicortical roots from base and at least sometimes from stem; young stems no more than slightly compressed, not keeled, lacking furfuraceous lines 7

7 Leaves narrowly ovate, base acute; inflorescence with 3-6 pairs of triads . . . . *4. P. nitens*
Leaves broadly ovate, base usually obtuse; inflorescence with to 12 or more pairs of triads . . . . *9. P. stelis*

8 Inflorescence umbellate or essentially so, at least when young, sometimes nearly sessile in leaf axil; flowers 6-merous . . . . *6. P. pycnostachya*
Inflorescence a (mostly pedunculate) spike or raceme; flowers (3-)4-merous 9

9 Members of a triad pair often separated along elongating inflorescence axis; triad bract broad, investing base of fruit; leaves thin, often mucronate or emarginate . . . . *3. P. guyanensis*
Inflorescence axis not elongating, triads strictly paired; triad bract not investing base of fruit; leaves coriaceous, not mucronate or emarginate 10

10 Inflorescence axillary only; bracts and bracteoles distinct, persistent on lower part of spikes; anthers alternatingly 2- and 4-loculate; dioecious, only male ones known . . . . *1. P. coarctata*
Inflorescence axillary and often in terminal compound clusters; bracts and bracteoles forming a deciduous cupule, leaving lower part of peduncle naked; anthers 2-loculate only; flowers bisexual . . . . *2. P. disjectifolia*

1. **Phthirusa coarctata** A.C. Sm., Lloydia 2: 175. 1939. Type: Guyana, A.C. Smith 2204 (holotype F, isotypes K, NY, U, MO).

Rather small plant, twigs slightly furfuraceous when very young, becoming smooth. Leaves (sub)opposite; petiole to 0.6 cm long; blade somewhat coriaceous, (ob)lanceolate to somewhat obovate, to 6 x 3 cm, apex obtuse, base acute; venation pinnate. Inflorescence solitary, axillary, to 3 cm long when in fruit; triads sessile; bracts and bracteoles persistent or deciduous below fruits. Petals and stamens 4, dimorphic, nearly sessile anthers alternatingly 2- and 4-loculate; style 0.1 cm long, slightly clavate. Fruiting spike becoming at least 3.5 cm long. Fruit yellow, ellipsoid, 0.5 x 0.35 mm, apex blunt.

Distribution: A rare species, perhaps known only from two specimens (GU: 2).

Specimens examined: Guyana: the type; Rupununi Distr., Massara, Graham 254 (K).

Note: See note under *Phthirusa disjectifolia*.

2. **Phthirusa disjectifolia** (Rizzini) Kuijt, Proc. Kon. Ned. Akad. Wetensch., Biol. Chem. Geol. Phys. Med. Sci. 93: 114. 1990. – *Furarium disjectifolium* Rizzini, Rodriguésia 18-19 (30-31): 155, t. 24. 1956. Type: Brazil, Amazonas, Rio Negro, Murça Pires 223 (holotype IAN).

Sparsely branched plants, basal epicortical roots few and slender; internodes terete. Leaves coriaceous, ovate to broadly lanceolate, to 8 x 4 cm; venation obscure. Inflorescence axillary and/or in elongated, terminal compound groups, each to 3 cm long, very short-pedunculate. Flowers 4-merous, to 0.15 cm long, in paired, sessile triads; stamens slightly dimorphic, anthers 2-loculate. Fruit ellipsoid, 0.3 x 0.2 cm, apex blunt.

Distribution: Venezuela (Amazonas, Bolívar), Guyana, and adjacent Brazil; 14 collections studied, 9 from the Guianas (GU: 9).

Selected specimens: Guyana: Basin of Shodikar Cr., Essequibo tributary, A.C. Smith 2849 (K); Upper Demerara R., Jenman 4122 (K); Mabura Hill and vicinity, Pipoly *et al.* 8935 (NY).

Note: Superficially somewhat similar to *P. coarctata*. The latter species, however, has distinct bracts and bracteoles, while those of *P. disjectifolia* are fused in cupulate fashion. Additionally, these 'cupules' in *P. disjectifolia* fall away as a unit after the fruits are gone, leaving the lower spike naked, while the bracts and bracteoles of *P. coarctata* persist. In the latter, 2-loculate and 4-loculate anthers alternate, but *P. disjectifolia* seems to have 2-loculate anthers only.

3. **Phthirusa guyanensis** Eichler in Mart., Fl. Bras. 5(2): 64. 1868. Type: Guyana, Rupununi R., Ri. Schomburgk 1602 (holotype B not seen, see F Negative # 11791).

Phthirusa perforata Rizzini, Revista Fac. Agron. (Maracay) 8: 91. 1975. Type: Venezuela, Steyermark & Bunting 102749 (holotype RB, isotype MO).
Phthirusa calloso-albida Rizzini, Ernstia 24: 15. 1984. Type: Venezuela, Huber & Tillett 5550 (holotype RB, isotype VEN).

Sparsely branched plants with slender epicortical roots at base; internodes to 7 cm long, terete; all young and inflorescence parts lightly brown- to grey-furfuraceous (especially in inflorescences). Leaves irregularly paired; petiole 0.5-1.5 cm long; blade oblanceolate to narrowly elliptic, to 8 x 3.5 cm, apex rounded to very slightly emarginate, minutely apiculate, base acutely tapering to distinct petiole; venation pinnate, with 2 or 4 strong lateral veins from near base, midrib running into apex, its lower side furfuraceous. Dioecious, male inflorescence not seen. Female inflorescence at least 2 cm at anthesis, indeterminate and apparently only axillary (no compound inflorescences seen), elongating to 7 cm in fruit, members of all but distal pair of triads becoming much separated from each other; triads sessile, 2-4 pairs per inflorescence; bracts and bracteoles forming a united greyish-furfuraceous cupule. Female flowers 4-merous, < 0.2 cm long; calyculus coarsely lacerate; petals scarcely or not dimorphic, ca. 0.1 cm long; sterile anthers not discernible; style clavate, 0.1 cm, stigma capitate. Fruit ellipsoid, 0.4 x 0.3 cm, pale orange, base invested by fused bracts.

Distribution: Venezuela, Guianas, and northern Brazil; ca. 15 collections studied (GU: 8).

Selected specimens: Guyana: Kaieteur Plateau, savanna, Maguire & Fanshawe 23278 (K, NY, U); Potaro-Siparuni Region, Kaieteur National Park, between airstrip and escarpment, Gillespie 949 (LEA, US).

Notes: Van Dam (Fl. Guianas, Suppl. 3: 185. 2002) reports that she has seen the type specimen in B. Its original name is ‘Viscum guianensis Klotzsch’, a nomen nudum, published in Richard Schomburgk, Reisen Brit.-Guiana 3: 1161. 1848. R. Vogt (B, pers. comm.) informed me that he could not find this specimen.
See also the comments under *Cladocolea micrantha*, which is often similar in general habit.

4. **Phthirusa nitens** (Mart.) Eichler in Mart., Fl. Bras. 5(2): 59. 1868. – *Loranthus nitens* Mart. in Schult. & Schult. f., Syst. Veg. 7: 150. 1829. – *Passovia nitens* (Mart.) Tiegh., Bull. Soc. Bot. France 42: 172. 1895, as ‘Passowia’. Type: Brazil, Amazonas, Martius s.n. (holotype M).

Slender plants with few basal epicortical roots; stems terete, sparsely branched. Petiole ca. 5 mm long; blade narrowly ovate, to 7 x 2 cm, apex attenuate, base acute. Inflorescence solitary in leaf axils but also in small, terminal compound clusters, each to 7 cm long, with 3-6 pairs of triads; triad peduncle 0.3-0.4 cm long, spreading; bracts and bracteoles slender, acute, persistent. Dioecious. Flowers 0.5-0.6 cm long (male). Fruit not known.

Distribution: Venezuela (Bolívar), Guyana and northern Brazil; 8 collections studied (GU: 4).

Specimens examined: Guyana: Mackenzie Mine area, E of Ituni, Cowan 39254 (US); margins of Upper Abary R., Maas *et al.* 5438 (MO, U); Upper Demerara-Berbice Region, 24 km from Ituni along Ituni-Kwakwani road, Gillespie *et al.* 3055 (LEA, US); Cuyuni-Mazaruni Region, savanna 2-3 km W of Maipuri Falls, Karowrieng R., Gillespie *et al.* 2876 (LEA, US).

Note: A rarely collected species which might conceivably be an unusually slender or shade form of *P. stelis*.

5. **Phthirusa podoptera** (Cham. & Schltdl.) Kuijt, Taxon 43: 198. 1994. – *Loranthus podopterus* Cham. & Schltdl., Linnaea 3: 218. 1828. Type: Brazil, Alagoa, Gardner 1330 (neotype P, isoneotype NY, designated by Kuijt in Kuijt & Kellogg 1996: 47).

   Loranthus pterygopus Mart. in Schult. & Schult. f., Syst. Veg. 7: 155. 1829. – *Struthanthus pterygopus* (Mart.) Mart., Flora 13: 105. 1830. Type: Brazil, Minas Gerais, Martius s.n. (lectotype M, designated by Kuijt 1994: 198).

Leafy plants, sparsely branched with long, slender, slightly compressed shoots soon becoming terete. Leaves thin, ovate, to 9 x 4 cm, apex acute, base of blade obtuse; venation pinnate. Inflorescence to 9 cm long, with 5-8 pairs of triads; inflorescence peduncle to 2.5 cm long, with 2 thin wings together 0.3-0.4 cm wide; all flowers sessile directly on inflorescence axis; small, acute, persistent bracts and bracteoles. Flowers bisexual. Petals and stamens strongly dimorphic, lower series of anthers 4-, upper series 2-loculate, filament deeply excavated; style straight, somewhat clavate. Fruit obovoid-ellipsoid, 0.7 x 0.4 cm, yellow.

Distribution: Northern Brazil, and most likely present in the adjacent Guianas; ca. 15 collections studied, not yet reported from the Guianas.

Note: An unmistakable species because of its flattened inflorescence peduncle, which is not present in any other S American member of the family.

6. **Phthirusa pycnostachya** Eichler in Mart., Fl. Bras. 5(2): 62. 1868. – *Passovia pycnostachya* (Eichler) Tiegh., Bull. Soc. Bot. France 42: 172. 1895, as 'Passowia'. Type: French Guiana, Poiteau s.n. (holotype P). – Fig. 7

Phthirusa monetaria Sandwith, Bull. Misc. Inform. Kew 1932: 227. 1932. Type: Guyana, Sandwith 313 (holotype K, isotype U).
Struthanthus umbellatus Kuijt, Proc. Kon. Ned. Akad. Wetensch., Biol. Chem. Geol. Phys. Med. Sci. 93: 126. 1990. Type: French Guiana, Cremers 9424 (holotype CAY, isotype LEA).

Leafy plants of medium size, basal epicortical roots present; internodes to 3.5 cm long, slightly quadrangular becoming terete, greyish with irregular brown lenticels. Petiole 0.2-0.3 cm thick and long, massive, distinct; blade coriaceous, broadly elliptic to nearly orbicular, to 7 x 6 cm, base obtuse; a number of strong palmate veins nearly reaching the apex. Inflorescences 1-3 per axil, umbellate to racemose, to 2.5 cm (rarely to 8 cm; Cremers *et al.* 12770, LEA) long; peduncle about 1 cm or nearly absent, terete, without basal crater, beyond which 4-20 pairs of triads crowded so closely together that often no space is left between them on the rather fleshy axis, rarely elongating somewhat; triad peduncle 0.2-0.3 cm long; bracts and bracteoles small, persistent, together forming a 3-alveolate structure bearing 3 sessile, somewhat sunken flowers. Flowers bisexual. Mature flower buds 0.2 cm long. Flowers 6-merous; petals red, strongly dimorphic; upper filament buttresses with deep lateral cavities formed by anthers of lower series, pollen sacs 4, oblong and of equal size, parallel to each other on a broad head with extremely slender attachment to filament (shorter stamen series), or on massive, excavated filament (longer series), connectival horns blunt; style hexagonal, clavately expanded upwards, somewhat constricted in middle, stigma undifferentiated. Fruit ovoid-elliptical, ca. 0.35 x 0.25 cm, blue and yellow, apex truncate.

Distribution: Guyana and French Guiana, not yet reported from Suriname, but to be expected there; 12 collections studied (GU: 4; FG: 8).

Specimens examined: Guyana: Moraballi Cr., near Bartica, Sandwith 313 (K, NY, U); Essequibo R., near Bartica, Jenman 2534 (K, NY); Willems Timber Concession, Polak *et al.* 551 (LEA, U); Cuyuni-Mazaruni Region, Kartabo, Willems Timber Concession, Hahn *et al.*

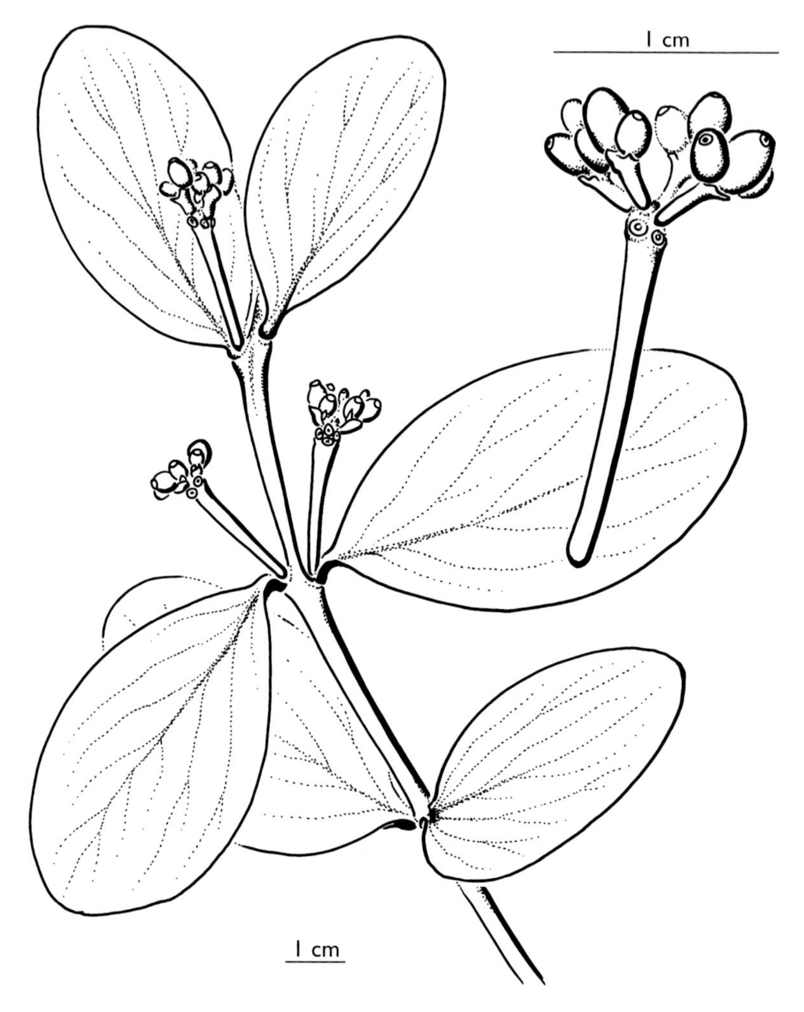

Fig. 7. *Phthirusa pycnostachya* Eichler: habit of young fruiting plant, with detail of infructescence (reproduced from Kuijt, Proc. Kon. Ned. Akad. Wetensch. 93, 1990).

5143 (LEA, US). French Guiana: the type; Vicinity of Saül, Eaux Claires, on slope, Prance *et al.* 30630 (LEA, NY); Vicinity of Saül, Route de Bélizon, < 500 m S of Eaux Claires, Mori *et al.* 23969 (LEA, NY); Mts. de la Trinité, bassin de la Mana, Cremers 12770 (CAY, LEA); R. Comté, à proximité de la route N2, Billiet & Jadin 1250 (BR); Mont Chauve, Cremers *et al.* 14867 (CAY, LEA); Route du Tour de l'Île de Cayenne RN 2, Savane du PK 16, Cremers 9424 (CAY, LEA); SE Cayenne, pont de la RN 2 sur la Comté, de Granville 3228 (CAY, LEA).

N o t e : There is a possibility that two different species are involved, the western plants virtually lacking a peduncle.

7. **Phthirusa pyrifolia** (Kunth) Eichler in Mart., Fl. Bras. 5(2): 63, t. 17. 1868. – *Loranthus pyrifolius* Kunth in Humb., Bonpl. & Kunth, Nov. Gen. Sp. ed. qu. 3: 441. 1820. – *Passovia pyrifolia* (Kunth) Tiegh., Bull. Soc. Bot. France 42: 172. 1895, as 'Passowia'. Type: Colombia, Cauca, Humboldt & Bonpland 1872 (holotype P).
– Fig. 8

Loranthus perrottetii DC., Prodr. 4: 292. 1830. Type: French Guiana, Perrottet s.n. (holotype G-DC, isotype US).

Epicortical roots from base of plant only, often profuse; stems compressed and keeled when young, often 0.6 cm or more wide just below nodes, completely and finely furfuraceous or with distict furfuraceous lines separated by green epidermis. Petiole distinct, ca. 1 cm long, somewhat compressed and keeled; blade dark green, shiny when fresh, rather thin, broadly lanceolate, ca. 10(-18) x 4-5 cm but sometimes larger, apex somewhat attenuate or even mucronate, base obtuse to truncate; venation pinnate. Inflorescence mostly 1 per foliar axil, sometimes clustered but not in terminal groups, 6-8 cm long or more with at least 14-20 triads on short peduncles or quite sessile; axis olive green to furfuraceous, of rough texture, terete. Flowers sessile, laterals with 2 prophylls each forming a small, dentate floral cup; petals dark wine-red to pale red, 0.15 cm long; stamens sessile, with brief, fleshy filament. Fruit ellipsoid, ca. 0.8 x 0.5 cm, rather blunt, at maturity bright orange red with yellowish apex and dark, purple-green base; embryo with large primary haustorium at maturity.

D i s t r i b u t i o n : Southern Mexico (Chiapas) through C America, Jamaica and most of tropical S America, including eastern Peru and Bolivia; probably the most widespread of neotropical Loranthaceae; many hundreds of collections studied, ca. 35 from the Guianas (GU: 10; SU: 6; FG: 17).

S e l e c t e d   s p e c i m e n s : Guyana: Rupununi Distr., Kuyuwini Landing, Kuyuwini R., Jansen-Jacobs *et al.* 3087 (LEA, U); Basin of Rupununi R., near mouth of Charwair Cr., A.C. Smith 2386 (K, MO, P). Suriname: Paramaribo, Eliselaan, Proctor 4725 (K); Saramacca R. headwaters, trail to Coppenam R., rear of village of Pakka Pakka, Maguire 23973 (K, MO). French Guiana: Roche Koutoo, Bassin du Haut-Marouini, de Granville *et al.* 9330 (CAY, LEA); Haut Marouini, (Itany), Antecum-Pata (Malavate), right bank, Sastre *et al.* 3828 (P).

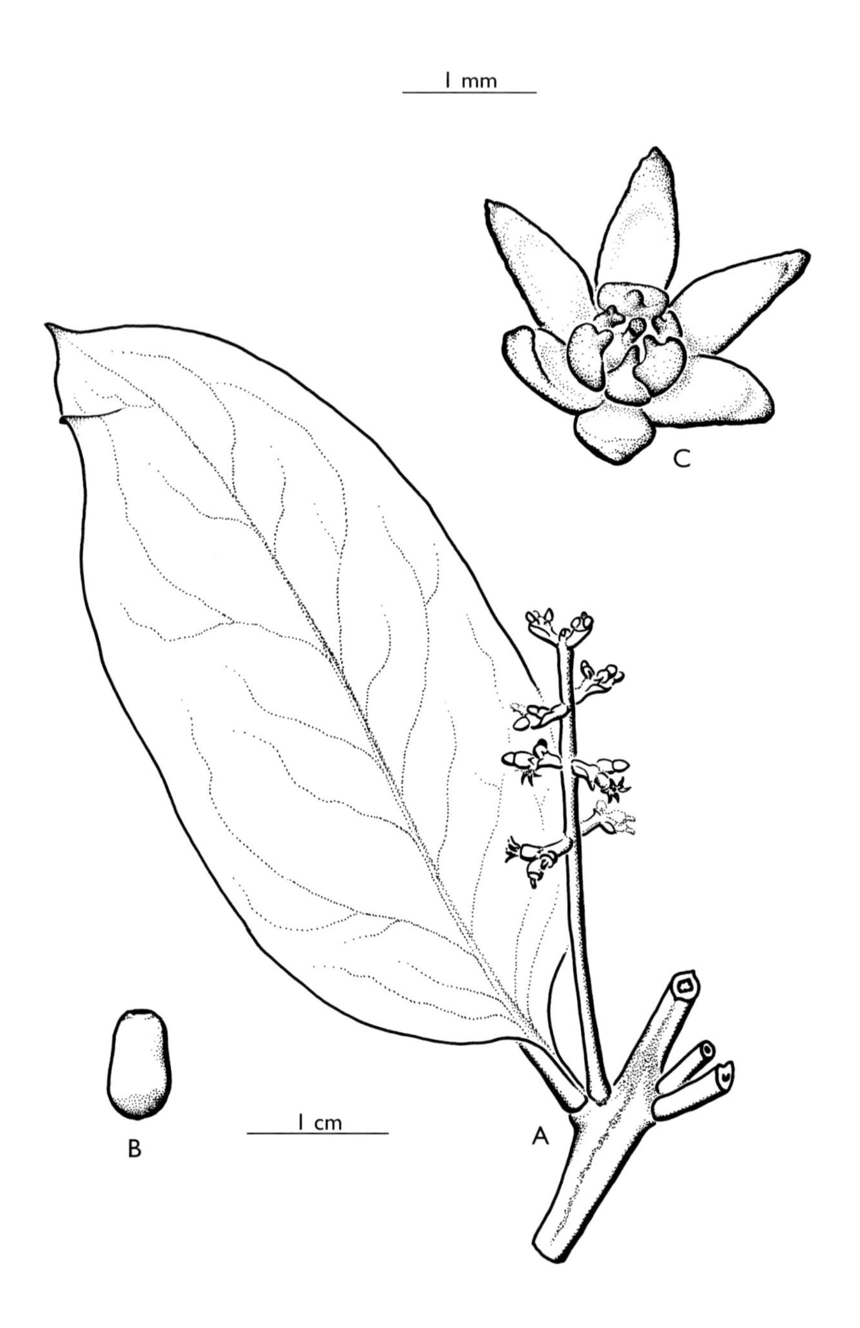

Fig. 8. *Phthirusa pyrifolia* (Kunth) Eichler: A, habit; B, mature fruit; C, flower (A-C, reproduced from Kuijt, Fl. of Ecuador 24, 1986).

Note: An occasional extremely vigorous plant of this species may be found with leaves to 20 x 10 cm.

8. **Phthirusa rufa** (Mart.) Eichler in Mart., Fl. Bras. 5(2): 61. 1868. – *Loranthus rufus* Mart. in Schult. & Schult. f., Syst. Veg. 7: 130. 1829. – *Struthanthus rufus* (Mart.) Mart., Flora 13: 105. 1830. Type: Brazil, Amazonas, Martius s.n. (holotype M).

   Phthirusa squamulosa Eichler in Mart., Fl. Bras. 5(2): 62. 1868. Type: Suriname, Bergendal, Wullschlaegel 227 (lectotype BR, designated by Kuijt 1994: 194).

Stout, sparsely branched plants with basal epicortical roots; young twigs (including inflorescences) strongly furfuraceous, light cinnamon to chocolate brown; stems terete. Petiole to 0.3 cm thick and 2 cm long; blade paler green beneath when fresh, leathery, ovate-lanceolate, to 18 x 5 cm, apex acute, base truncate or nearly so; venation pinnate, midvein very strongly marked. Dioecious. Inflorescence solitary and simple when in axils of several pairs of upper leaves, dark brown, often grading into a terminal compound raceme with several-10 lateral pairs of racemes and a terminal one, individual racemes mostly to 8 cm long, occasionally to 15 cm long, peduncle 1.5 cm long followed by some 8-10(-20) pairs of triads; triad peduncle perpendicular to axis, 0.2-0.5 cm long. Flowers sessile, yellowish to bright red, 0.5 cm long, 0.1 cm of which is ovary; anthers yellow, dimorphic, connective short and blunt; style 0.3 cm long, in female flower upper $^{2}/_{3}$ thickened in clavate fashion but terminating in truncate, capitate stigma; in male flower style thickened in middle. Fruit purplish, ellipsoid, 1 x 0.6 cm.

Distribution: Guyana, Suriname and northern Brazil; 50+ collections studied, at least 40 of which from the Guianas, to be expected in French Guiana (GU: 30+; SU: 3).

Selected specimens: Guyana: East Demerara Region, Yarowkabra settlement & Forestry Station, 6 km ESE of station, Pipoly *et al.* 7373 (LEA, P, US); Upper Demerara-Berbice Region, 5 km E of Rockstone, Linden-Rockstone road, Pipoly *et al.* 9616 (NY); Upper Takutu-Upper Essequibo Region, along road N from Karanambo between compound and dam reservoir, McDowell *et al.*1871 LEA, US); Mabura Hill, Ekuk compartment, Jansen-Jacobs *et al.* 1990 (LEA, U); Demerara-Mahaica Region, along Linden Hwy, 6 km S of Kuru-Kuru Cr., Hoffman *et al.* 710 (LEA, US); Cuyuni-Mazaruni Region, Pakaraima Mts., 0.5-1 km SE of Imbaimadai towards Partang R. mouth, Hoffman *et al.* 1675 (LEA, US); Upper Demerara-Berbice Region,

Linden-Soesdyke Hwy, between Dora and Maibia Cr., Pipoly *et al.* 9681 (LEA, US). Suriname: Vicinity of Zanderij, 45-50 km S of Paramaribo, near Matta, Mori *et al.* 8328 (NY); Coesewijne Savanna, Lindeman & Mennega *et al.* 184 (U, VEN).

9. **Phthirusa stelis** (L.) Kuijt, Taxon 43: 193. 1994. – *Loranthus stelis* L., Sp. Pl. ed. 2. 473. 1762. – *Struthanthus stelis* (L.) Blume in Schult. & Schult. f., Syst. Veg. 7: 1731. 1830. Type: Panama, Hammel 3298 (neotype MO, isoneotype LEA, designated by Kuijt in Kuijt & Kellogg 1996: 50). – Fig. 9

Loranthus retroflexus Ruiz & Pav., Fl. Peruv. Chil. 3: 49, t. 279, f. a. 1802. – *Phthirusa retroflexa* (Ruiz & Pav.) Kuijt, Brittonia 32: 521. 1981 ('1980'). Type: Peru, Pavon s.n. (isotype MO).
Loranthus aduncus G. Mey., Prim. Flora Esseq. 149. 1818. – *Phthirusa adunca* (G. Mey.) Maguire in Maguire *et al.*, Bull. Torrey Bot. Club 75: 301. 1948. Type: not designated.
Loranthus theobromae Willd. in Schult. & Schult. f., Syst. Veg. 7: 132. 1829. – *Phthirusa theobromae* (Willd.) Eichler in Mart., Fl. Bras. 5(2): 56. 1868. – *Passovia theobromae* (Willd.) Tiegh., Bull. Soc. Bot. France 42: 172. 1895, as 'Passowia'. Type: Brazil, Martius s.n. (holotype M).
Loranthus erythrocarpus Mart. in Schult. & Schult. f., Syst. Veg. 7: 138. 1829. – *Phthirusa erythrocarpa* (Mart.) Eichler in Mart., Fl. Bras. 5(2): 58. 1868. – *Passovia erythrocarpa* (Mart.) Tiegh., Bull. Soc. Bot. France 42: 172. 1895, as 'Passowia'. Type: Brazil, Amazonas, Martius s.n. (holotype M).
Phthirusa polystachya Eichler in Mart., Fl. Bras. 5(2): 57. 1868. – *Passovia polystachya* (Eichler) Tiegh., Bull. Soc. Bot. France 42: 172. 1895, as 'Passowia'. Type: Brazil, Pará, Spruce 1018 (holotype P).
Phthirusa seitzii Krug & Urb. in Urb., Bot. Jahrb. Syst. 24: 16. 1897. Type: Tobago, Eggers 5521 (lectotype P, here designated).
Phthirusa angulata K. Krause, Recueil Trav. Bot. Neérl. 22: 344. 1925. Type: Suriname, Coppename R., Gonggrijp & Stahel 1104 (U), syn. nov.

Scandent plants, epicortical roots at base and occasionally from stems; young internodes slightly compressed, becoming terete, glabrous or essentially so. Petiole about 1 cm long; blade often yellowish or light green, broadly ovate, to 14 x 7.5 cm, apex usually attenuate and base obtuse. Dioecious. Inflorescence both terminal and axillary, in the former case always and in the latter case often compound, consisting of a terminal raceme plus 2-6 lower, lateral racemes each with about 12 pairs of evenly spaced triads, compound inflorescence to 15 cm; triad peduncle to 0.2 cm, terminating in a bract and 2 bracteoles of about equal size, these persisting, acute, subtending 1 sessile flower each. Flower bud

Fig. 9. *Phthirusa stelis* (L.) Kuijt: habit and mature fruit (reproduced from Kuijt, Fl. of Ecuador 24, 1986).

to 0.6 cm long, female rather acute and narrow, male blunt. Petals whitish yellow to light red; aborted organs well differentiated in flowers of opposite sex; anther connective forming small, terminal horn. Fruit brick red to orange, ellipsoid, to 1 x 0.5 cm.

Distribution: As defined here, *P. stelis* ranges from Costa Rica through most of lowland tropical S America, on the east side of the Andes south to Bolivia; many hundreds of collections studied, at least 120 from the Guianas (GU: 25+; SU: 6; FG: 30+).

Selected specimens: Guyana: Upper Demerara-Berbice Region, Linden-Soesdyke Hwy between Dora and Maibia Cr., Pipoly *et al.* 9694 (P); Mazaruni Forestry Station, Maguire *et al.* 23578 (NY, K). Suriname: Paramaribo, Essedstraat, Proctor 4728 (MO, U); Wilhelmina Mts., Zuid R., Kayser Airstrip, 45 km above confluence with Lucie R., Irwin *et al.* 57608 (MO). French Guiana: Bassin de l'Approuague, Mts. des Nouragues, bords de cours d'eau, Larpin 870 (P); Saül, Jardin Botanique de l'Orstom, Oldeman B-3913 (P).

10. **Phthirusa stenophylla** Eichler in Mart., Fl. Bras. 5(2): 60, t. 19, f. 6. 1868. Type: Brazil, Spruce 3307 (holotype P, isotypes BR, F, G, NY).

Slender, sparsely branched plants, with few epicortical roots from stem; internodes to 8 cm long, terete, glabrous. Petiole 0.1-0.2 cm long; blade thin, very narrowly lanceolate, to 4 x 0.8 cm, apex acute, apiculate, base acute; obscure venation except for midrib. Inflorescence to 1.5 cm, peduncle 0.2-0.5 cm, followed by 1 or 2 pairs of triads; triad peduncle ca. 0.4 cm; bracts and bracteoles short and blunt, persistent. Fruit greyish green with white dots, ellipsoid, 0.6 x 0.3 cm.

Note: Inclusion of this species is based on a single collection (Henkel *et al.* 762 (LEA)), providing the above diagnosis which closely parallels the protologue, in which the stamens are shown to be of the same type as that in, for example, *P. pyrifolia*. It is clearly an extremely rare species, of uncertain distribution; its type originated in southern Amazonas. The species may be confused with the very narrow-leaved form of *Cladocolea micrantha* which, however, has furfuraceous stems and 4-merous flowers.

11. **Phthirusa trichodes** Rizzini, Revista Fac. Agron. (Maracay) 8: 92. 1975. Type: Venezuela, Lara, Steyermark *et al.* 103648 (holotype RB "masculus", isotype MO).

Very small and delicate species, haustorial saddle 1.5 cm in diam., lacking epicortical roots; young growth finely pubescent; internodes terete, mostly less than 1 cm long. Petiole 0.1-0.2 cm long; blade very thin, lanceolate to 1.5 x 0.8 cm; delicate, basally pinnate venation.

Inflorescence ca. 0.5 cm long, to 1 cm in fruit, with extremely slender, pubescent peduncle and 2 sessile triads at tip, in umbellate fashion. Flowers apparently bisexual; petals 4, ca. 0.1 cm long; stamens isomorphic, filaments short, terete. Fruit ellipsoid, 0.4 x 0.2 cm, with small, smooth, erect calyculus.

Distribution: Known from Venezuela (Lara, Portuguesa, and Táchira) and, more recently, from 1 collection in Guyana (GU: 1).

Specimen examined: Guyana: Upper Takutu-Upper Essequibo Region, Rewa R., forest 4 km W of camp, elev. 180 m, dense forest and brown sand, Clarke 3729 (LEA, US).

Note: The only known *Phthirusa* species with true indument, *P. trichodes* is a puzzling entity which may not be correctly placed in the genus.

6. **PSITTACANTHUS** Mart., Flora 13: 106. 1830.
Type: P. americanus (L.) Mart. (Loranthus americanus L.)

Often large, leafy, glabrous parasites, primary haustorium massive; epicortical roots mostly lacking; stems terete or quadrangular. Leaves usually paired, sometimes whorled; blades often fleshy or leathery. Inflorescences indeterminate racemes or umbels of triads or dyads, former in terminal or subterminal positions, latter axillary. Flowers large, mostly pedicellate (sessile in *P. cucullaris*), subtended by 1 bract or bracteole each; bisexual, 6-merous, usually brightly colored in red or red and yellow; petals and stamens dimorphic, rarely isomorphic; petals long, slender, fused to varying length, bearing epipetalous stamens consisting of slender filaments, anthers dorsifixed, mostly versatile, 4-loculate. Fruit a large berry; seed consisting mostly of a massive, fleshy green seedling with 2 or more large, fleshy cotyledons, endosperm lacking. (x = 8, 10).

Distribution: Mostly continental, northwestern Mexico (including central Baja California) to Argentina; a large genus of ca. 120 species; in the Guianas 12 species.

Note: A possibly additional species not included separately below is present in the Matoroni R. area of French Guiana, where flowers were collected from the ground (Mori *et al.* 25307, LEA; see the citation under *P. clusiifolius*). Its petals are nearly twice as long (12-13 cm) as those of the latter species, the stigma much larger and strikingly papillate; and the thick, rapidly tapering filament is flanked by tufts of long, red hairs – features missing in *P. clusiifolius*. The plant may represent an undescribed species.

LITERATURE

Kuijt, J. 1983. Status of the genera Aetanthus and Psathyranthus (Loranthaceae). Candollea 38: 661-672.

Kuijt, J. & D. Lye. 2005. Gross xylem structure of the interface of Psittacanthus ramiflorus (Loranthaceae) with its host and with a hyperparasite. Bot. J. Linn. Soc. 147: 197-201.

Kuijt, J. (under review). Monograph of Psittacanthus (Loranthaceae). Syst. Bot. Monogr.

KEY TO THE SPECIES

1 Inflorescence a double triad (umbel of 2 triads) or raceme of dyads or triads 2
  Inflorescence an umbel (or near-umbel) of 4 triads 9

2 Flowers sessile at tip of triad peduncle; ovary completely immersed in an elongated, tubular cupule; bract broad, foliar . . . . *4. P. cucullaris*
  Flowers pedicellate; ovary at least partly emergent from cupule or fused bracteoles; bract not foliar 3

3 Leaves sessile, base cordate, clasping the stem . . . . *3. P. cordatus*
  Leaves petiolate, base not cordate 4

4 Inflorescence bearing dyads 5
  Inflorescence bearing triads 6

5 Inflorescences placed at branch tips, each a raceme; leaves similar on both sides; shoot tip aborting; stamens isomorphic, filaments 4-5 mm long, thick . . . . *2. P. clusiifolius*
  Inflorescences axillary along shoot, each a double dyad; stamens dimorphic, overlapping for half their length; lower leaf surface brownish; shoot tips not aborting; filaments 7-8 mm long, thin . . . . *11. P. redactus*

6 Internodes angular when young; bud apex acute; stamens nearly isomorphic, with red hairs . . . . *8. P. melinonii*
  Internodes terete when young; bud apex obtuse to rounded; stamens dimorphic, lacking hairs 7

7 Buds and inflorescence branches very stout; inflorescence a terminal raceme of triads; petals fleshy, to 6.3 cm long, with smooth margins . . . . *1. P. acinarius*
  Buds and inflorescence branches not stout; inflorescences axillary, each a double dyad; petals not fleshy, < 4.5 cm long, with crenate margins . . . 8

8 Leaf apex acute; buds no more than very short-hairy; midrib prominent, reaching leaf apex . . . . *6. P. grandifolius*
Leaf apex rounded; buds densely brown-hairy; midrib not prominent, not reaching leaf apex . . . . *7. P. lasianthus*

9 Leaf apex acute; petal tips black, usually recurved somewhat even in bud, with granular surface, basal ligules hairy . . . . *9. P. peronopetalus*
Leaf apex rounded; petal tips not black or recurved, glabrous, basal ligules absent 10

10 Petals to 7.5 cm long, filiform; filaments 3.5-4 cm long . . . *12. P. robustus*
Petals < 4.5 cm long, not filiform; filaments < 1.5 cm long 11

11 Leaves broadly ovate to lanceolate, venation evident; buds strongly curved; anthers lacking hairs . . . . *5. P. eucalyptifolius*
Leaves ovate, flat and leathery, venation rather obscure; buds straight; anthers with long red hairs . . . . *10. P. plagiophyllus*

1. **Psittacanthus acinarius** (Mart.) Mart., Flora 13: 108. 1830. – *Loranthus acinarius* Mart. in Schultes & Schultes f., Syst. Veg. 7: 130. 1829. Type: Brazil, Piauí, Martius s.n. (holotype M). – Fig. 10

Psittacanthus corynocephalus Eichler in Mart., Fl. Bras. 5(2): 36. 1868. Type: Brazil, Bahia, Blanchet 3160 (lectotype BR, isolectotypes F, G, P, designated by Kuijt 1994: 195), syn. nov.

Stout, brittle plants; leafy stems terete or nearly so, to 1 cm thick. Leaves paired; short-petioled to nearly sessile; blade very fleshy, irregularly orbicular-obovate or asymmetrical, to 11 x 11 cm; venation pinnate. Inflorescence terminal and in nearby axils; triad peduncle and pedicels stout; bracteoles blunt and short; cupule heavy, investing basal $1/4$ of barrel-shaped ovary 0.7 x 0.6 cm; calyculus smooth or irregularly dentate. Flower bud 0.5 cm thick below, 1.5-2 cm long and to 0.7 cm thick, with clavate, blunt tip; petals orange red with yellowish or light green tip, very fleshy, to 6.3 cm long, spongy "ligule" at tip encasing stigma and tips of upper stamen series; stamens dimorphic, filament attached to petal ca. 4.3 cm above base, stout, 1-1.5 cm long, anthers light yellow, 0.7-0.8 cm long, blunt, overlapping slightly; style straight, stigma 0.15 cm thick, finely papillate, placed well above anthers. Fruit purple with pink base, ovate, 2 x 1 cm.

Distribution: Colombia, Venezuela, French Guiana, Ecuador, Peru, Bolivia and adjacent Brazil, one disjunct population in southern Costa Rica (Oso Peninsula); locally often infrequent; ca. 50 collections studied, 2 from French Guiana, but to be expected in Suriname and Guyana (FG: 2).

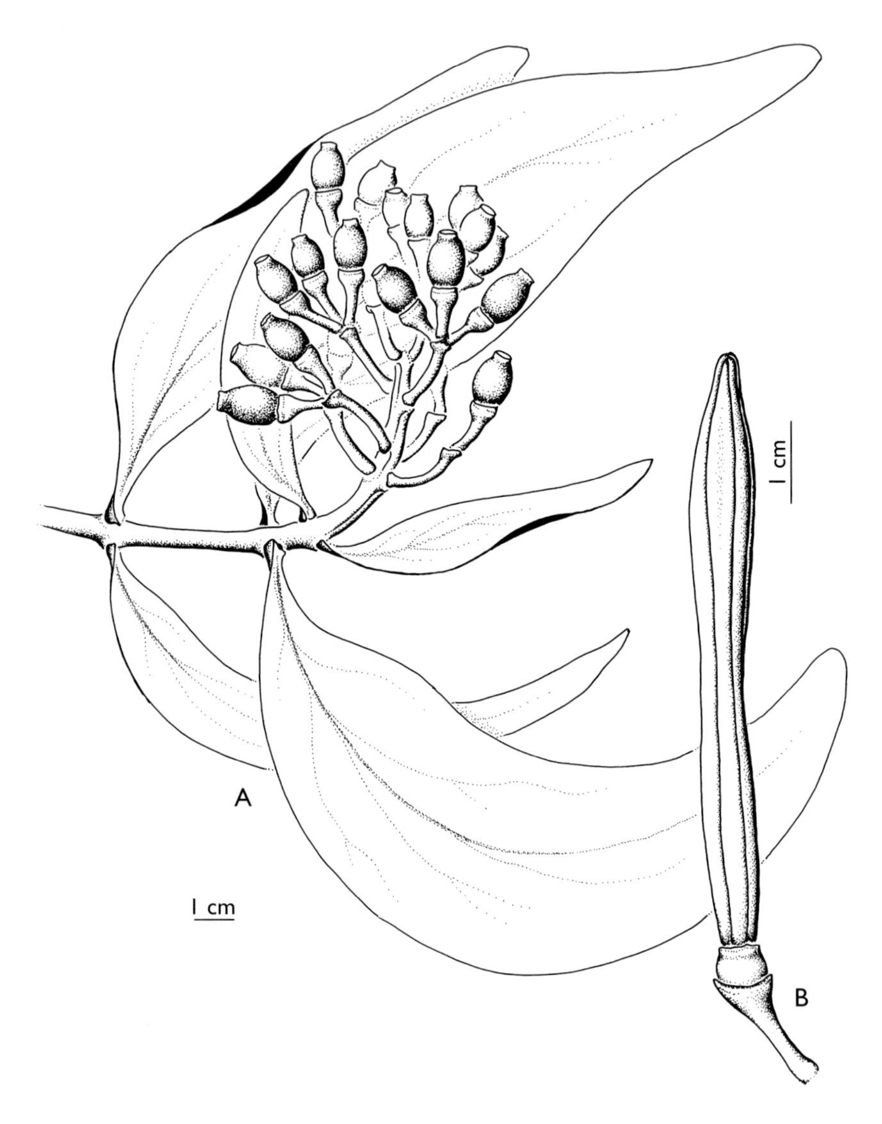

Fig. 10. *Psittacanthus acinarius* (Mart.) Mart.: A, fruiting habit; B, flower bud (based on: A, Gentry *et al.* 2245; B, Grández & Ruiz 2631).

Specimens examined: French Guiana: Regina Region, eastern plateau of Mts. Tortue, 11 km WNW of Approuague R., Feuillet *et al.* 9966 (LEA); Saül, La Fumée Mt., La Fumée Ouest, 200-300 m elev., Mori *et al.* 18896 (NY).

2. **Psittacanthus clusiifolius** Eichler in Mart., Fl. Bras. 5(2): 30. 1868, as 'clusiaefolius'. Type: Brazil, Spruce 1890 (lectotype M, isolectotype K, designated by Kuijt 1994: 195). – Fig. 11

Stems terete, internodes fairly stout, to 6 cm long. Petiole stout, to 1 cm long; blade leathery, elliptical, to 12 x 5 cm, apex rounded, base acute to nearly obtuse; evident pinnate venation. Inflorescence apparently not truly terminal, but rather axillary to upper 2 leaves; axis 2-3 cm long, bearing 3-4 pairs of dyads; dyad peduncle to 0.5 cm long, with inconspicuous bract, pedicels slightly shorter; cupule small; calyculus smooth. Flower bud 8.5 cm long (12 cm in Mori *et al.* 25307), including 0.5 x 0.2 cm cylindrical ovary, straight, clavate, about 0.1 cm thick below, gradually expanding to 0.3 cm 3/4 up, then narrowing to 0.2 cm again and expanding to 0.8 x 0.3 cm thick tip with rounded apex; petals scarlet in- and outside (as filaments and style), but apex bright yellow outside, reflexing only at expanded tip; stamens isomorphic, filament fairly stout, 0.4-0.5 cm long, anthers nearly basifixed, erect, yellow, 0.6 cm long, pollen sacs showing a rather moniliform structure; style nearly as long as bud, with conspicuous, capitate stigma. Fruit purplish black, ellipsoid, 1.1 x 0.7 cm, pedicel purple, calyculus inconspicuous.

Distribution: Brazilian Amazonas; 15 collections studied, 4 of which from the Guianas (GU: 1; SU: 1; FG: 2).

Specimens examined: Guyana: Upper Mazaruni Basin, Mt. Ayanganna, along N side, 900 m elev., Tillett *et al.* 45878 (NY). Suriname: Gonini R., Versteeg 241 (U). French Guiana: Bassin de l'Approuague, Savane Roche de Virginie, on inselberg, Cremers *et al.* 15291 (CAY, LEA); Mataroni R., base camp just above second rapids, Mori *et al.* 25307 (LEA).

3. **Psittacanthus cordatus** (Hoffmanns. ex Schult. & Schult. f.) Blume in Schult. & Schult. f., Syst. Veg. 7: 1730. 1830. – *Loranthus cordatus* Hoffmanns. ex Schult. & Schult. f., Syst. Veg. 7: 128. 1829. Type: Brazil, without collector (holotype 6976 B-W, not seen). – Fig. 12

Branches to 2 m long, internodes terete, to 6 cm long. Leaves sessile to amplexicaul, paired; blade leathery, long-cordate or lingulate, to 10 x 5 cm, apex rounded, margin smooth, thin, yellow; venation palmate. Inflorescence mostly terminal, often with smaller units in nearby leaf axils, at least lower portions very finely granular, with 2-3 pairs of triads; inflorescence peduncle and triad peduncle about 1 cm long, latter

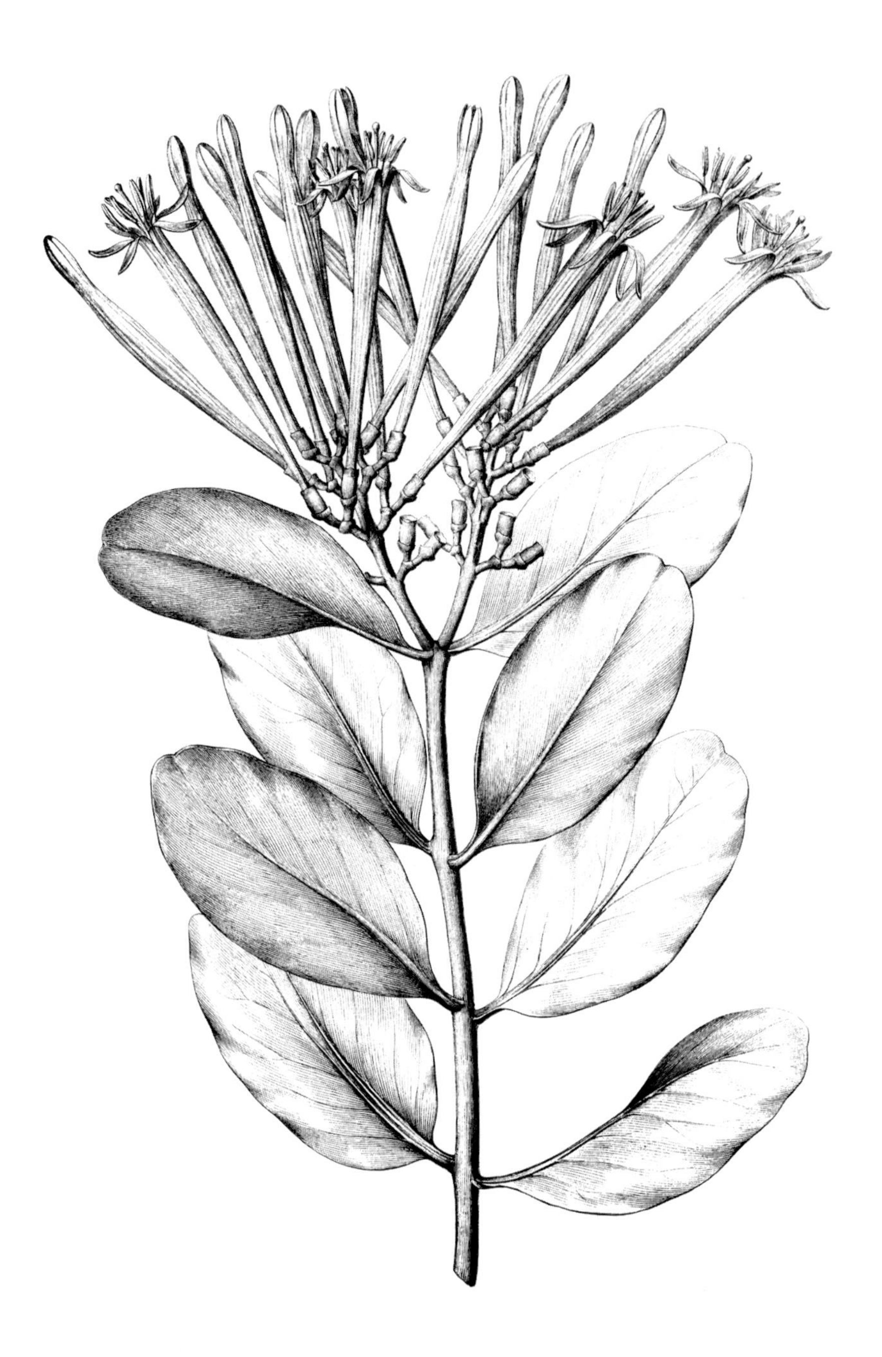

Fig. 11. *Psittacanthus clusiifolius* Eichler: flowering habit (reproduced from Eichler, Flora Brasiliensis 5(2), 1868).

Fig. 12. *Psittacanthus cordatus* (Hoffmanns. ex Schult. & Schult.f.) Blume: flowering habit (reproduced from Eichler, Flora Brasiliensis 5(2), 1868).

with narrowly or broadly oblong bract of 0.4-0.5 cm long; pedicel ca. 0.5 cm long; cupule 0.5 x 0.4 cm, reddish brown, obscuring entire ovary or nearly so at anthesis, margin splitting in fruit; cayculus smooth. Flower bud straight to slightly curved, 0.25 cm thick, 1 cm long tip expanded to 0.4 cm thick, apex blunt; petals 4-4.5 cm long, recurved in upper 1/2 or 1/3 only, light orange-red; stamens dimorphic, filaments 1.4 cm long, anthers 0.45 cm long, rather blunt at both ends; style straight, 4 cm long, stigma weakly differentiated. Fruit 1 x 0.8 cm, calyculus prominent, erect.

Distribution: Venezuela, Guyana, south to Paraguay and north-central Argentina, Amazonian Bolivia; not yet reported from Suriname and French Guiana; 50+ collections studied (GU: 7).

Specimens examined: Guyana: Kanuku Mts., Rupununi R., Bush Mouth near Witaru Falls, Jansen-Jacobs *et al.* 99 (LEA, U, US); Rupununi Distr., Chaakoitou, near Mountain Point, just S of Kanuku Mts., Maas *et al.* 4073 (LEA, U); Rupununi Distr., Dadanawa, along Rupununi R., Jansen-Jacobs *et al.* 2034 (LEA, U), 3908 (LEA, U); Rupununi Distr., Manari, Takatu R., Jansen-Jacobs *et al.* 4768 (LEA, U); Upper Takutu-Upper Essequibo Region, S Rupununi Savanna, Ororkar bus, SE of Aishalton, Henkel *et al.* 3729 (LEA, US); Dadanawa Ranch, near ranch compound, Clarke *et al.* 1650 (LEA, US).

4. **Psittacanthus cucullaris** (Lam.) Blume in Schult. & Schult. f., Syst. Veg. 7: 1730. 1830. – *Loranthus cucullaris* Lam., J. Hist. Nat. 1: 444, t. 23. 1792. Type: French Guiana, Leblond s.n. (P, not located); Suriname, Nickerie, area of Kabalebo Dam Project, Lindeman & Görts *et al.* 651 (neotype US, isoneotypes F, NY, U, here designated). – Fig. 13

Loranthus falcifrons Mart. in Schult. & Schult. f., Syst. Veg. 7: 129. 1829. – *Psittacanthus falcifrons* (Mart.) Mart., Flora 13: 107. 1830. Type: Brazil, Amazonas, Martius s.n. (holotype M).

Stems more or less terete, nodes slightly swollen. Petiole to 5 mm long; blade variable in size and shape, ovate to lanceolate, often asymmetrical, to 20 x 7 cm, base cuneately tapering into flat; venation palmate, with 3-5 large, obvious veins running most of the length of blade. Inflorescence terminal and in nearby axils, 1-3 per axil, each a short raceme of 2-6 triads; triad peduncle 1-1.5 cm long, terminating in a conspicuous, clasping bract which is variable in size, sometimes (perhaps only in the Guianas area) being large and broadly foliaceous; flowers sessile; cupule bright red, usually narrow, thick-walled, about 1 cm long, margin of cupule with 1 or

Fig. 13. *Psittacanthus cucullaris* (Lam.) Blume: habit of flowering branch (reproduced from Kuijt, Fl. of Ecuador 24, 1986).

several acute teeth, cupule completely hiding ovary of 0.15 cm wide and 0.1 cm long, which is almost doubled in length by erect, smooth calyculus. Flower bud straight and slender, clavate, 0.3 cm wide in terminal 1 cm, the acute tip showing 6 free slender tips, 3 long and 3 short ones; petals slender, to 4 cm long, yellow with basal $^1/_3$ red to reddish-orange, tip somewhat greenish, petals strongly recurving from point of filament insertion (about 2.5 cm from base); stamens dimorphic, slender filaments

1.2-1.5 cm, anthers narrow, 0.25-0.3 cm, with some very long, red hairs on back, the 2 series slightly overlapping; style straight and slender, stigma finely papillate, oblong, capitate, lodged against petal tips beyond anthers. Fruit blackish, ovoid, about 1.5 cm long, basal half sheathed in enlarged cupule; embryo with (3) 4 cotyledons.

Distribution: Apparently a common species in the western Amazonian basin from Colombia, Venezuela, the Guianas, Brazil, Ecuador, Peru, Bolivia, and in the Pacific lowlands of Colombia and Ecuador; 1 disjunct population in Costa Rica (Oso Peninsula); 120+ collections studied, of which >20 in the Guianas (GU: 1, SU: 18; FG: 4).

Selected specimens: Guyana: Berbice, Ro. Schomburgk ser. I, 344 (K). Suriname: Kayser Mts., 330 km SSW of Paramaribo, 9 km SW of Kayser Airstrip, vicinity of old GMD Camp II, Mori *et al.* 8609 (LEA, NY); Saramacca R. headwaters, Flat Rock, 7 hours above Posoegronoe, Maguire 24920 (NY); Bank of Saramacca R., above Base Camp of line to Ebbatop, just above Paka paka, Florschütz *et al.* 1625 (U); Upper Coppename R., Boon 1146 (U); Upper Litanie R., Versteeg 407 (U). French Guiana: Primary forest near Zidockville, Haut Oyapock, Prévost *et al.* 997 (LEA); Saut Pakusili, primary forest, Grenand 1470 (LEA).

Note: The species is unique in *Psittacanthus* for the unusual development of the cupule (easily mistaken for the ovary and calyculus) which has been morphologically controversial (Kuijt 1981), and possibly unique for the 3 flowers, all of which are sessile.

5. **Psittacanthus eucalyptifolius** (Kunth) G. Don, Gen. Hist. 3: 417. 1834. – *Loranthus eucalyptifolius* Kunth in Humb., Bonpl. & Kunth, Nov. Gen. Sp. ed. qu. 3: 432. 1820. Type: Venezuela, Prov. Caracas, Humboldt & Bonpland 766 (holotype P-Bonpl.). – Fig. 14

   Psittacanthus collum-cygni Eichler in Mart., Fl. Bras. 5(2): 35. 1868. Type: Brazil, Pará, Spruce 137 (lectotype M, designated by Kuijt 1994: 195), syn. nov.

Stems terete, internodes to 4 cm. Leaves paired; petiole rather stout, 0.3-0.4 cm long; blade broadly ovate to lanceolate, to 12 x 6 cm, with rather tapering but ultimately rounded tip and cuneate base; venation pinnate. Innovations and inflorescence with craterlike scar, latter axillary usually with 4 triads in a early umbellate cluster; inflorescence peduncle 1.5-3 cm; triad peduncle ca. 1 cm, with short, wide, blunt bract; flower pedicel slightly longer; cupule investing only base of ovary, latter nearly cylindrical but widening somewhat at smooth calyculus, 0.5 x 0.3 cm.

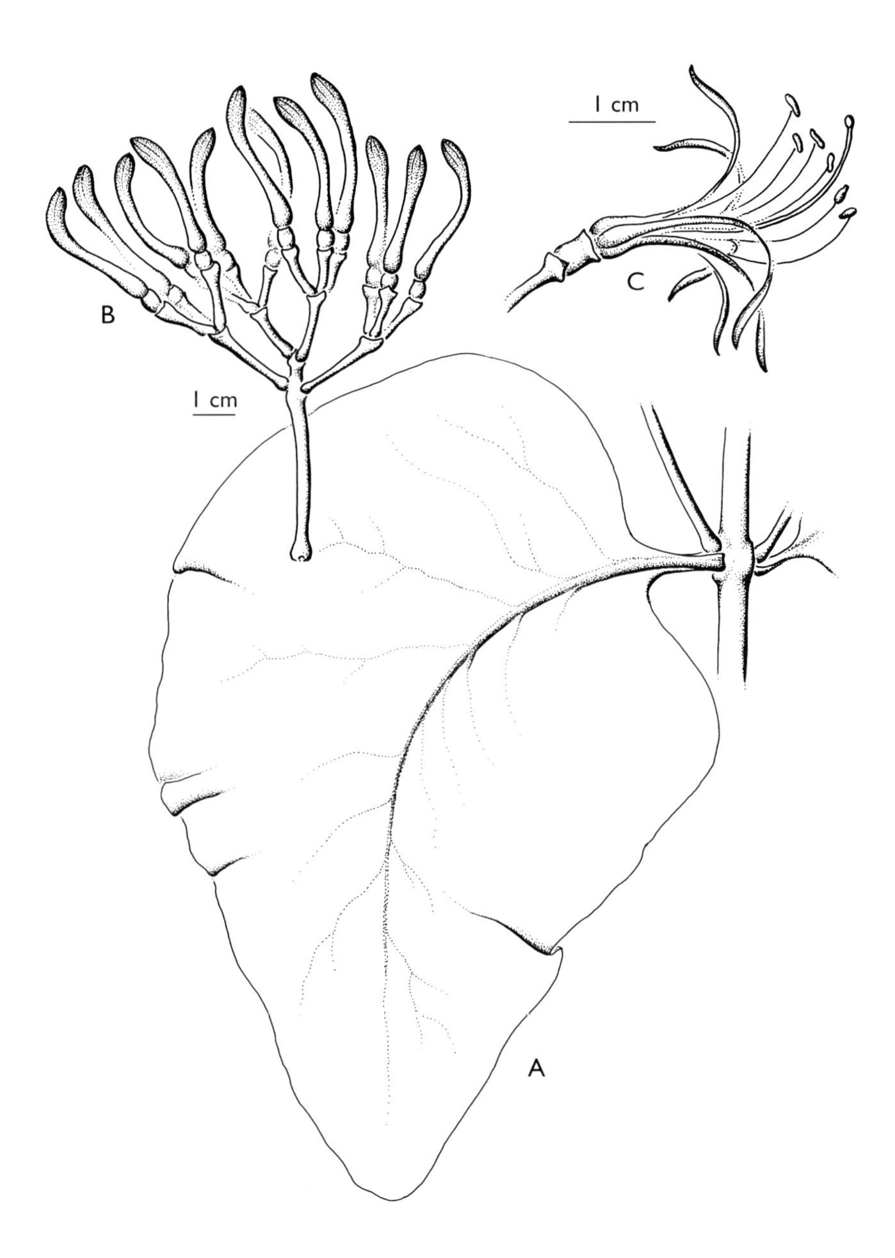

Fig. 14. *Psittacanthus eucalyptifolius* (Kunth) G. Don: A, typical, large leaf; B, inflorescence; C, expanded flower (based on: Berg *et al*. BC 614).

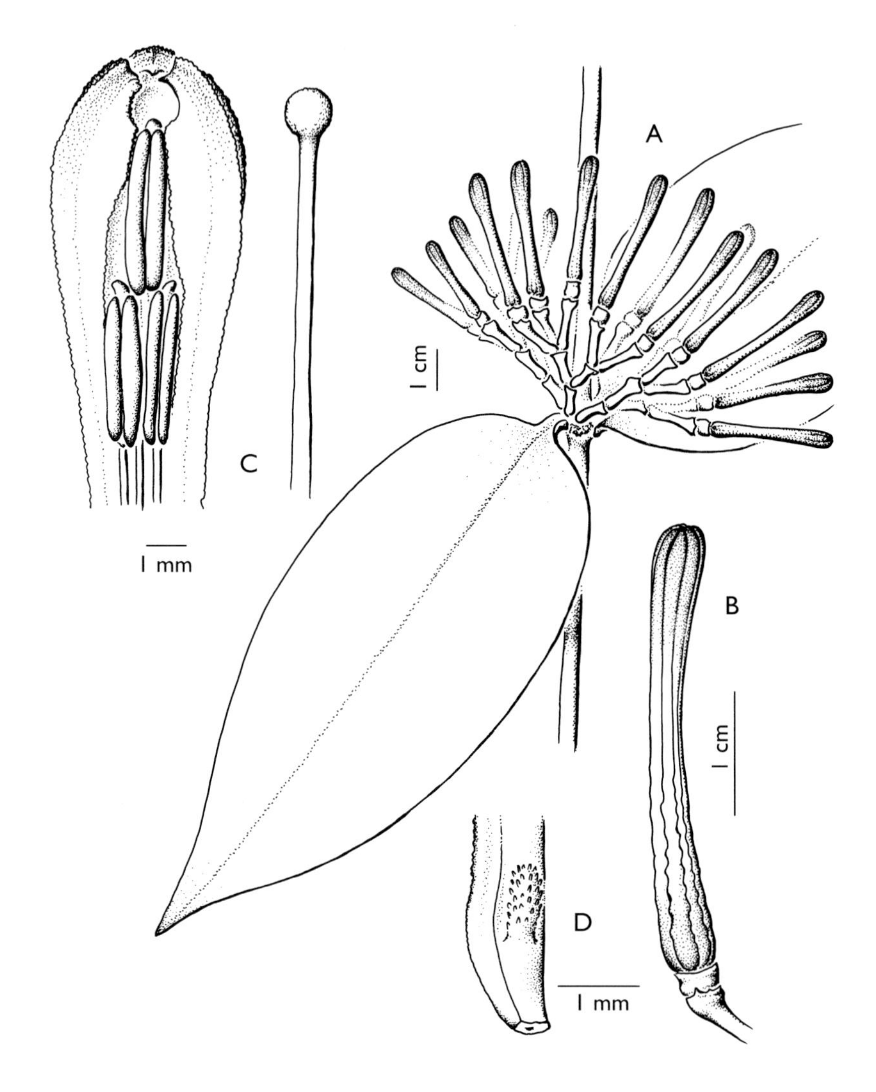

Fig. 15. *Psittacanthus grandifolius* (Mart.) Mart.: A, habit; B, flower bud; C, dissected flower bud apex and style; D basal ligule (based on: A, Cid *et al.* 1455; B-D, Cremers 9899).

Flower bud ca. 3.5 cm long, strongly sigmoid, 0.2-0.3 cm thick, tip expanded to 0.3 cm, rather blunt; petals with elaborate terminal ligule, said to be yellow, opening for at least 3/4; inside of petal "puckered" when dry; filament ca.1.2 cm long, anthers short and thick, 0.3 x 0.15 cm. Fruit 1.2 x 0.8 cm, with well developed calyculus.

Distribution: Upper Amazonian forests of Venezuela, Brazil, to E Peru; ca. 30 collections studied (SU: 3).

Specimens examined: Suriname: Sipaliwini Savanna area on Brazilian frontier, Oldenburger *et al.* 374, 816 (U); above Sipaliwini, Kamp XXI, savanna, Rombouts 510 (U).

6. **Psittacanthus grandifolius** (Mart.) Mart., Flora 13: 108. 1830. – *Loranthus grandifolius* Mart. in Schult. & Schult. f., Syst. Veg. 7: 124. 1829. Type: Brazil, Amazonas, Martius s.n. (holotype M). – Fig. 15

Rather large plants with terete stems. Petiole stout, 0.2-0.6 cm long; blade leathery, ovate, to 17 x 7 cm, apex acute, sometimes attenuate, base obtuse to nearly truncate, lower surface dull, upper one shiny; venation pinnate, lower midrib very prominent. Inflorescence axillary, mostly an umbel of 2 stout triads (rarely a raceme of 4 triads), ramifications purplish red; inflorescence peduncle to 1 cm, triad peduncles and pedicels ca. 0.5 cm long; cupules well-formed, investing base of ovary, 0.3 cm long. Flower bud ca. straight and clavate; petals red below with yellowish orange apex, to 4.5 cm long, base of tube 0.2-0.3 cm, narrowing 2/3 up, tip truncate, to 0.3 cm wide, surface evenly short-hairy, basal ligule yellowish, short-hairy, 0.2 cm long; stamens dimorphic, attached where tube narrowest, filaments ca. 1 cm long, anthers 0.3 cm long, not overlapping; style as long as petals, pink, stigma capitate. Fruit black, ovoid, 1 x 0.6 cm.

Distribution: Guyana, French Guiana and adjacent Brazilian Amazonas; infrequently collected; 6 collections studied (GU: 1 FG: 5).

Specimens examined: Guyana: Upper Takutu-Upper Essequibo Region, Kassikaityu R., 3-4 km N of landing at terminus of rail from Kuyuwini R., 240 m elev., Clarke 4819 (US). French Guiana: Oyapock, island near Saut Maripa, Prévost 1360 (CAY, LEA); Bassin de l'Oyapock, Cr. Gabaret, between mouth and Cr. Mérignan, Cremers 9899 (CAY, LEA, U, US); Banks of Oyapock, Saut Maripa, Jacquemin 2119 (CAY, LEA); Mt. de Kaw, trail to caves of Kaw, 33 km E of Roura on the Mt. de Kaw road, Mori *et al.* 25215 (LEA, NY); Piste de Degrad Lalanne-Mt. de Kaw, Feuillet 3594 (CAY, LEA, US).

7. **Psittacanthus lasianthus** Sandwith, Bull. Misc. Inform. Kew 1939: 18. 1939. Type: Guyana, Sandwith 1366 (holotype K). – Fig. 16

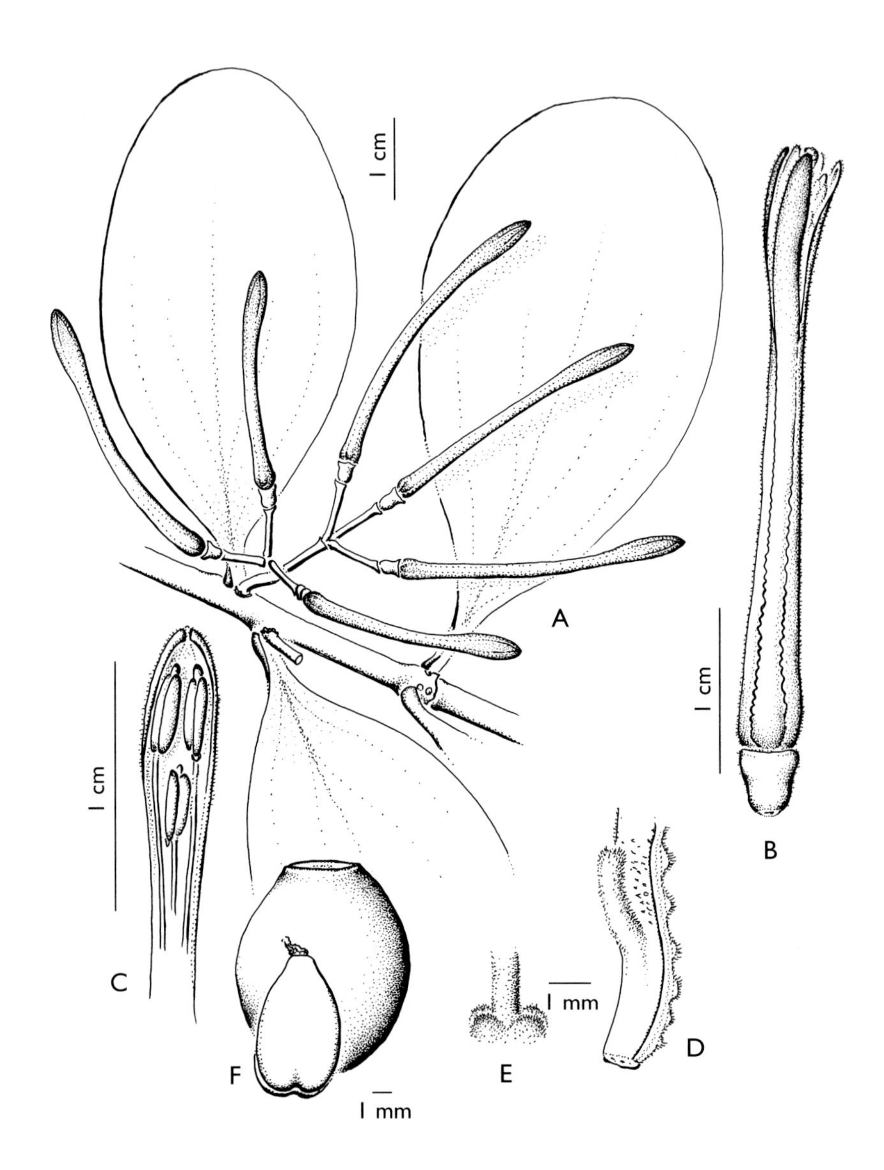

Fig. 16. *Psittacanthus lasianthus* Sandwith: A, habit; B, flower bud; C, dissected flower bud apex; D, basal ligule; E, stylar base and nectary; F, fruit and dicotylous embryo (based on: A-B, Stergios & Aymard 7670; C-E, Stergios *et al.* 8251; F, Lance *et al.* 66).

A rather small-leaved plant, stem stout, internodes short, 1.5-3 cm long, terete. Petiole massive, to 1 cm long; blade fleshy, glaucous when young, broadly ovate to nearly orbicular or somewhat spathulate, to 8 x 5 cm; venation obscure. Inflorescence an axillary 2-rayed umbel, red, base invested by small corky crater; peduncle of inflorescence and triad 0.7-0.9 cm; pedicel ca. 1 cm long, with conspicuous cupule; ovary 0.2-0.3 cm long and 0.4 cm wide, nearly glabrous; calyculus entire. Flower bud deep scarlet or red-orange at base, orange-yellow within, 4-4.5 cm long, base of flower tube dilated to 0.5 cm; petals furry-tomentose on back, petal tips each raised; stamens dimorphic, filaments attached 2 cm above base, 1.3-1.6 cm long, anthers 0.45 and 0.6 cm long, with conspicuous, smooth connectival horn 0.1-0.15 cm long, long, red hairs on back of anthers and on petals just above filament insertion; style reaching tips of upper anthers, stigma broadly clavate, smooth, nearly flat. Fruit becoming black, 1 x 0.8 cm; cotyledons 2.

Distribution: Venezuelan Amazonas and Guyana; 30+ collections studied, only 1 Venezuelan collection (Amazonas, Stergios & Aymard 7670, MO) (GU: ca. 32).

Selected specimens: Guyana: Mazaruni-Potaro Region, Upper Mazaruni R. Basin, N of Karowrieng R., Pakaraima Mts., Pipoly *et al.* 7763 (LEA, US); Potaro-Siparuni Region, Kaieteur Falls National Park, Hahn *et al.* 4048 (LEA, US).

Note: Indumentum is rare in *Psittacanthus*, and here (on the flower buds) consists of short hairs as well as dense clusters of longer, pointed, flat hairs; the latter are especially conspicuous on petal margins, and sometimes form transverse ridges on the petal backs.

8. **Psittacanthus melinonii** (Tiegh.) Engl. in Engl. & Prantl, Nat. Pflanzenfam. ed. 2. 16b: 180. 1935. – *Alveolina melinonii* Tiegh., Bull. Soc. Bot. France 42: 360. 1895, as ‘melinoni’. Type: French Guiana, Mélinon 145 (holotype P). – Fig. 17

Medium sized plants, young stems glabrous, somewhat angular, becoming terete. Leaves paired; petiole 1-1.5 cm long; blade coriaceous, obovate, to 10 x 5 cm, acutely tapering to petiole, margin brown when dry; venation obscure but pinnate. Inflorescence axillary only?, apparently on leafless older wood, consisting of a pair of triads; inflorescence peduncle 0.4 cm, triad peduncle and pedicels 0.5 cm long, slightly furfuraceous; bract and bracteoles fairly small. Flower bud straight and slender, ca. 7 cm long, of which cylindrical ovary is 0.4 cm

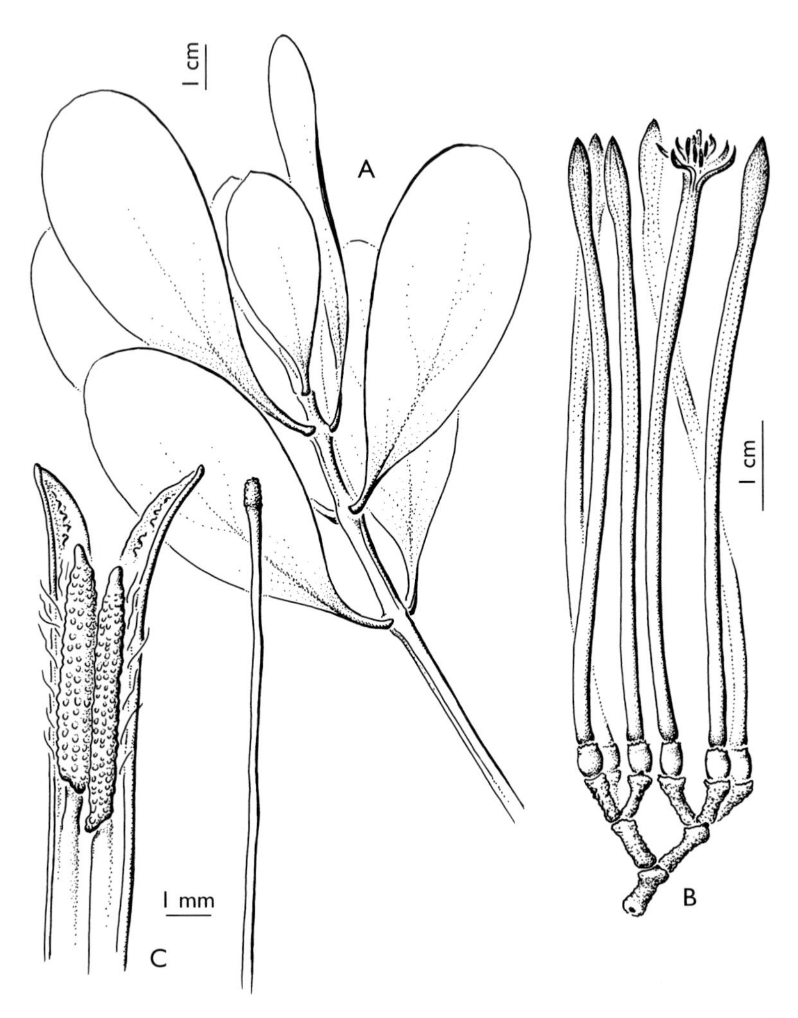

Fig. 17. *Psittacanthus melinonii* (Tiegh.) Engl.: A, habit; B, inflorescence; C, dissected flower bud apex (based on: Cremers 9872).

long and 0.15 cm thick above irregularly dentate calyculus, gradually expanding to 0.3 cm, with an indistinct neck 1 cm below acute tip; petals carmine red, ovary reddish violet; stamens nearly isomorphic, filament 1.5 cm long, inserted ca. 6 cm above petal base, anthers 0.6 cm long, acicular, with 4 rows of minute locules dehiscing separately, anther back with numerous long, unbranched, stiff, reddish hairs; style as long as flower bud, stigma somewhat clavate but essentially undifferentiated.

Distribution: French Guiana (endemic?) and possibly adjacent Brazilian Amapá; 2 collections studied (FG: 2).

Specimens examined: French Guiana: the type; Cr. Gabaret, Bassin de l'Oyapock, entre embouchure et Cr. Mérignan, Cremers 9872 (BR, LEA).

9. **Psittacanthus peronopetalus** Eichler in Mart., Fl. Bras. 5(2): 31. 1868. Type: Brazil, Amazonas, Spruce 1047 (lectotype M, isolectotype G, designated by Kuijt 1994: 196). – Fig. 18

Stems terete, internodes to 7 cm long. Leaves (sub)opposite; petiole short, 0.2-0.5 cm long, stout; blade narrowly ovate, to 15 x 6 cm, apex somewhat attenuate, base truncate or nearly so; venation pinnate but inconspicuous. Inflorescence (mostly) lateral, often clustered near branch tips, each a frequently long-stalked raceme, ca. 4 pairs of triads crowded near tip; inflorescence peduncle to 3 cm long, triad peduncle and pedicels 0.4-0.6 cm long; cupule investing ca. $^1/_3$-$^1/_2$ of ovary, entire; ovary 0.25 x 0.2 cm; calyculus irregularly crenate. Flower bud straight, 0.25 cm thick below, 0.2 cm at neck, tip 1 cm long, expanded to 0.3 cm thick, 0.2-0.3 cm of petal tips linear, recurving in bud, each terminating in a small, dark horn; petals vermillion, 4.2-4.5 cm inside and margin with sparse, short, thick hairs becoming dense at base of petal, where also a hair-fringed ligule of 0.15 cm long; filaments ca. 1 cm, dorsifixed in short sleeve of anther, and attached ca. $^2/_3$ up petal; style more or less straight, stigma clavate. Fruit ellipsoid, 1 x 0.5 cm, deep red.

Distribution: Known mostly from Brazilian Amazonas, but also reported from Amazonian Venezuela, Ecuador and Peru; ca. 40 collections studied, including 2 from the Guianas (GU: 1; FG: 1).

Specimens examined: Guyana: Cuyuni-Mazaruni Region, Ayanganna Plateau, 3 km N of Koatse R., 650-670 m elev., on Myrsine, Pipoly *et al.* 10700 (NY). French Guiana: Fleuve Oyapock, sur la crique Armontabo à environ 5.5 km en amont du Saut Canori, Oldeman B-1486 (U).

10. **Psittacanthus plagiophyllus** Eichler in Mart., Fl. Bras. 5(2): 37. 1868. Type: Brazil, Pará, Spruce 136 (lectotype M, designated by Kuijt 1994: 196). – Fig. 19

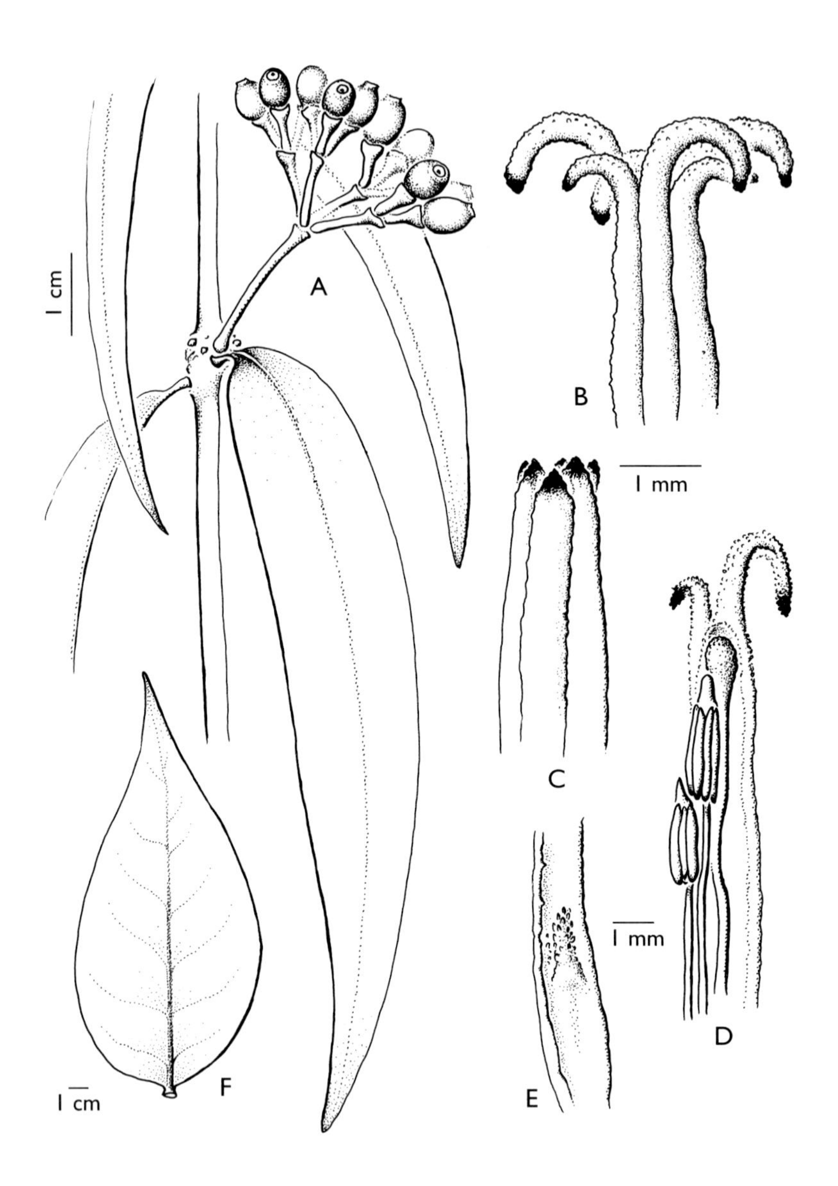

Fig. 18. *Psittacanthus peronopetalus* Eichler: A, fruiting habit; B, apex of flower bud, common type; C, apex of flower bud, the type with straight petal tips; D, dissected flower bud apex; E, basal ligule; F, leaf (based on: A, Kubitzki *et al.* P21702; B, D-E, Nelson 924; C, Wurdack & Adderley 43429; F, Spruce 1338).

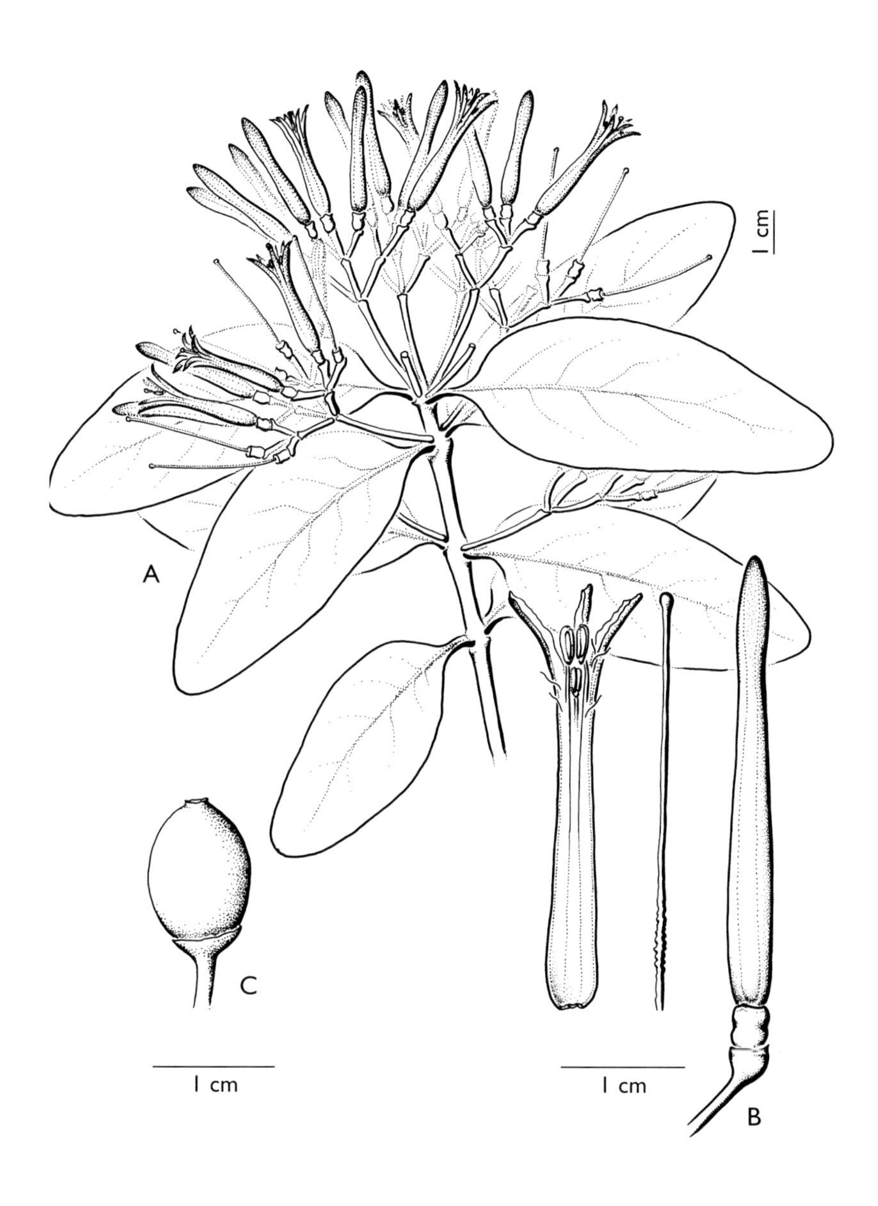

Fig. 19. *Psittacanthus plagiophyllus* Eichler: A, habit; B, dissected flower, style and flower bud; C, fruit (reproduced from Kuijt, Fl. of Ecuador 24, 1986).

Internodes fairly slender, terete, to 10 cm long. Petiole indistinct; blade stiff, greyish green, sometimes tending to be falcate, to 12 x 2 cm, base of blade acutely tapering; venation rather obscure, pinnate. Inflorescence in upper axils, especially crowded at stem tip, each an umbel of 4 triads; inflorescence peduncle to 3 cm long, triad peduncle 1-1.5 cm; bract rather small; pedicel to 0.7 cm long, expanding into a cupule 0.3 cm wide surrounding lower 1/4 of ovary, which is 0.3 cm long; calyculus irregulary but not deeply torn. Flower bud (minus ovary) 3 cm long, red below becoming yellow above, straight, 0.15 cm thick below, very slightly narrowing to a neck 1 cm below apex, end of bud 0.2 cm thick, apex acute; filaments attached at ca. 2 cm from petal base, ca. 0.3 cm long, anthers 0.2 cm, petals at that level bearing sparse hairs; stigma clavate, well differentiated. Fruit unknown.

Distribution: Colombia, Amazonian Venezuela, French Guiana, northern Brazil and Ecuador; ca. 20 collections studied, including 12 from the Guianas (GU: 6; SU: 4; FG: 2).

Selected specimens: Guyana: Bartica-Potaro road, near 14th milepost, Sandwith 1143 (K, U); Pomeroon R., Issororo Cr., For. Dept. D-275 (K); Mappo Cr., Berbice R., For. Dept. D-585 (K); Essequibo R., White Cr., Groete Cr., For. Dept. F-1783 (K); Bartica, For. Dept. F-3454 (K). Suriname: s.l., Hostmann 819 (K); Jodensavanne-Mapane Kreek area, oever Suriname R., Schulz 8699 (U); Forest Reserve, Sectie O, BW 4674 (U). French Guiana: Maroni, Penitencier St. Laurent, Mélinon s.n. (P); between Léandre and Roche Elisabeth, Lemée s.n. (P).

11. **Psittacanthus redactus** Rizzini, Rodriguésia 18-19(30-31): 145. 1956. Type: Brazil, Amapá, Oyapock, Black 49-8-445 (holotype IAN). – Fig. 20

Psittacanthus rufescens Rizzini in B. Maguire *et al.*, Mem. New York Bot. Gard. 29: 29. 1978. Brazil, Amapá, Pires *et al.* 51059 (holotype RB not seen, isotypes K, MO, U, US).

Medium-sized plant with fairly terete stems and 2 or 3 stout, unbranched epicortical roots from well above the host contact. Leaves decussate but members of a pair frequently separated 1-2 cm, small axillary corky craters present; petiole at least 0.5 cm long; blade leathery, lanceolate to slightly obovate, to 11 x 4 cm, apex blunt to slightly apiculate, base acute, cuneate; pinnate venation evident especially above. Inflorescence an umbel of 2 dyads; inflorescence peduncle less than 0.2 cm (sometimes very inconspicuous), dyad peduncle 0.1-0.2 cm, pedicels 0.3-0.4 cm long; bract and bracteoles scarcely recognizable; ovary 0.35 x 0.15 cm;

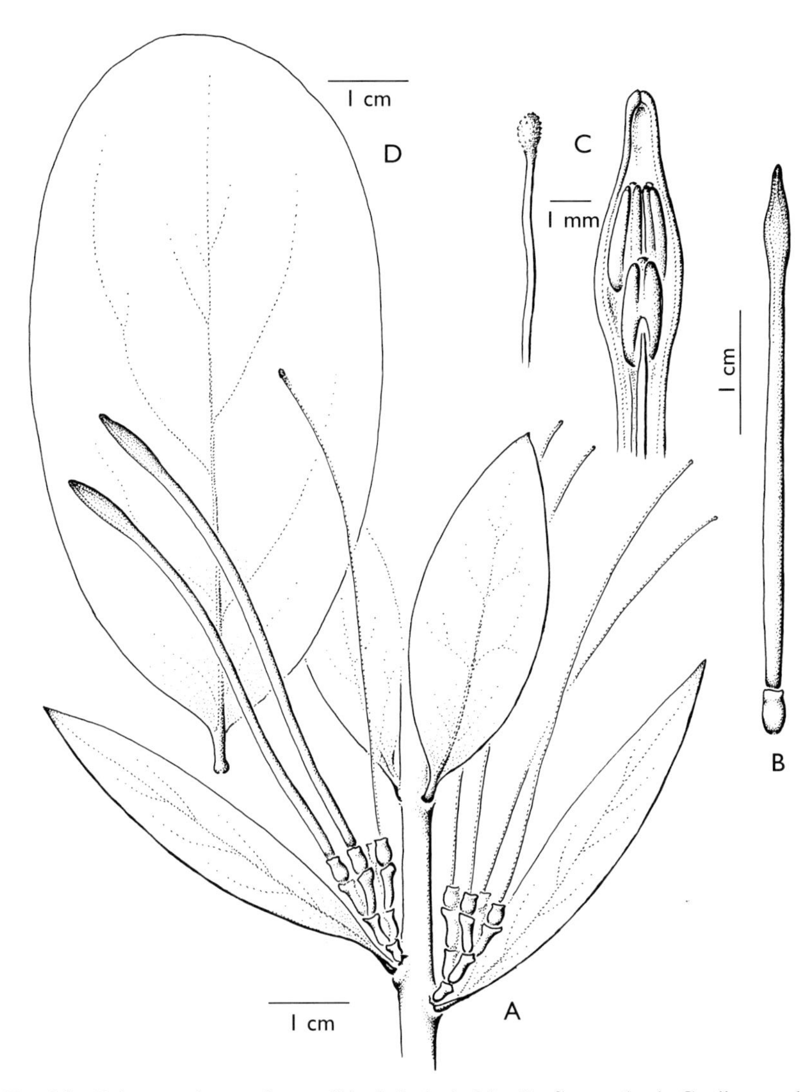

Fig. 20. *Psittacanthus redactus* Rizzini: A, habit; B, flower bud; C, dissected flower apex and style; D, leaf (based on: A, Santos 635; B, Pires *et al.* 51059).

calyculus more or less smooth. Flower bud slender, without constrictions or dilations, ca. 0.25 cm thick, gradually tapering to an acute apex; petals vermillion, 6.3 cm long; stamens dimorphic, yellow, filaments 0.5 and 0.8 cm long, attached 4.6 & 5 cm above petal base, anthers 0.4 x 0.05 cm, dorsifixed near the middle; style nearly as long as petals, stigma clavate. Fruit unknown.

Distribution: Brazilian Amapá and adjacent French Guiana; 7 collections studied, 4 of which from French Guiana (FG: 4).

Specimens examined: French Guiana: Mts. des Nouragues, Bassin de l'Aratraye, Sarthou 262 (CAY, LEA); Mts. Bakra, rocky savanna on granitic outcrop, 1.5 km W of Pic Coudreau, de Granville 4086 (CAY, LEA); Mts. Bakra- Pic Coudreau, de Granville 4113 (CAY, LEA); Mt. des Nouragues, Bassin de l'Approuague, savane-roche, Larpin 489 (CAY).

Note: This species is closely related to *P. lamprophyllus* Eichler, which has nearly identical flowers and inflorescences, but the leaves are extremely leathery and bicolored, the upper surface varnished strikingly, and with a sharply acute apex and massive, flat petiole. Whether *P. lamprophyllus*, as *P. redactus* and other related S American species, has epicortical roots remains to be seen. Such roots, with secondary haustoria, were first reported for *P. julianus* Rizzini (Ferrari, Revista Fac. Agron. (Maracay) 9: 93-95. 1976); other species in the genus are without epicortal roots. Curiously, in the two rooted specimens of *P. redactus* seen (de Granville 4113 and Cordeiro 281, the latter from Mpio. Barcelos, Amazonas, Brazil (LEA)), no secondary haustoria are visble on the epicortical roots, even though one on the former collection is 18 cm long.

12. **Psittacanthus robustus** (Mart.) Mart., Flora 13: 108. 1830. – *Loranthus robustus* Mart. in Schult. & Schult. f., Syst. Veg. 7: 125. 1829. Type: Brazil, Sâo Paulo, Martius s.n. (lectotype M, designated by Kuijt 1994: 196). – Fig. 21

Psittacanthus decipiens Eichler in Mart., Flora Bras. 5(2): 35. 1868. Type: Brazil, Amazonas, Spruce 1046 (lectotype M, isolectotypes P, NY, designated by Kuijt 1994: 195).
Psittacanthus intermedius Rizzini, Revista Fac. Agron. (Maracay) 8: 94, f. 10. 1975. Type: Venezuela, Amazonas, Cerro Yapacana, Steyermark & Bunting 103046 (holotype RB not seen, isotypes F, MO, NY).

Large, glabrous plants, stems somewhat angular when young, becoming terete. Leaves coriaceous, lance-elliptic to somewhat obovate, mostly to 9 x 5 cm; venation obscure. Inflorescence crowded in uppermost leaf axils, not terminal, each an umbel of 4 triads; inflorescence peduncle to 6 cm, triad peduncle to 1.5 cm, pedicels ca. 1 cm long; bract and braceoles rather small; cupule investing no more than 1/4 of ovary; calyculus ca. 0.2 cm long, shallowly and irregularly dentate. Flower bud to 8 cm long, straight and slender to very slightly arched, base of bud ca.

Fig. 21. *Psittacanthus robustus* (Mart.) Mart.: flowering habit (reproduced from Eichler, Flora Brasiliensis 5(2), 1868).

0.2 cm wide, narrowing to nearly half that, the 1-1.5 cm tip slightly expanded, apex narrowly rounded; petals and filaments extremely thin, almost filiform, filaments 3.5-4 cm long, implanted ca. 3 cm above base of petal, anthers 0.4 cm long, rounded at both ends.

Distribution: A common plant throughout Brazil, and in adjacent Venezuela, Colombia, and Bolivia; 70+ specimens studied, 2 from Guyana (GU: 2).

Specimens examined: Guyana: Upper Takutu-Upper Essequibo Region, Makarapan Mt., 1-2 km S of camp near base of S side of the mountain, 100 m elev., Clarke *et al.* 7029 (LEA, NY, US); Rupununi R., Monkey Pod Landing, SW of Mt. Makarapan, on Vochysiaceae, Maas *et al.* 7351 (F, LEA, MO, NY).

Note: *Psittacanthus robustus* is characterized by a profusion of hair-like, yellowish petals and stamens near the tips of leafy shoots, turning reddish in age.

7. **STRUTHANTHUS** Mart., Flora 13: 102. 1830, nom. cons.
   Type: S. syringifolius (Mart.) Mart. (Loranthus syringifolius Mart.)

Scandent, glabrous parasites with terete or quadrangular stems; epicortical roots both at base of plant and on twigs. Leaves paired. Dioecious. Inflorescences solitary or clustered, axillary, an indeterminate spike or raceme of paired triads. Flowers 6-merous, inconspicuous and small, usually greenish white; prominent staminodia and vestigial styles representing aborted organs in female and male flowers, respectively. Fruit an orange, reddish or blue berry; endosperm white, embryo bright green, dicotylous. ($x = 8$).

Distribution: A strictly continental genus, ranging from northern Mexico, to Bolivia and northern Argentina; uncertain number of species (see note below); in the Guianas 4 species.

Notes: *Struthanthus* is believed to be polyphyletic, separate species or species groups being derived from distinct branches of *Cladocolea*. The genus is generally of low-altitude preference. It has never been monographed; the number of species, while uncertain, is at least 50.
Much confusion has existed in the past with regard to the separation of *Struthanthus* from *Phthirusa*, but the great majority of species of the former genus are easily recognized in the male plant by its versatile anthers. Uncertainty persists vis-à-vis the boundary of *Struthanthus* with

Cladocolea, and with regard to certain small-flowered *Phthirusa* species like *P. guyanensis and P. trichodes*, which do not have the otherwise characteristic anther morphology of *Phthirusa*.
In the past, it was reported (Kuijt in Boggan *et al.*, Checkl. Pl. Guianas ed. 2. 1997) that *S. marginatus* (Desr.) Blume occurs in Guyana. This appears to be incorrect, it is a Brazilian species, only encountered remote from the Guianas.

KEY TO THE SPECIES

1 Inflorescence subtended by minute, broad and short, persistent basal scale leaves; youngest leaves often recurved, hook-shaped, prehensile . . . . *3. S. gracilis*
Inflorescence lacking basal scale leaves; youngest leaves neither hook-shaped nor prehensile 2

2 Leaf blade mostly < 2x as long as wide, obtuse at base and apex . . . . *4. S. syringifolius*
Leaf blade at least 2x as long as wide, leaf base (and normally also apex) acute 3

3 Leaves to 8 cm long, petiole to 1 cm long; triads 3 or 4 pairs . . . . *1. S. concinnus*
Leaves mostly to 3 cm long, petiole < 0.5 cm long; triads 1(2) pairs . . . . *2. S. dichotrianthus*

1. **Struthanthus concinnus** (Mart.) Mart., Flora 13: 105. 1830. – *Loranthus concinnus* Mart. in Schult. & Schult. f., Syst. Veg. 7: 150. 1829. Type: Brazil, Amazonas, Martius s.n. (M).

Plants sparsely branched, glabrous or minutely papillose; bearing occasional epicortical roots; internodes terete or slightly compressed. Petiole to 1 cm long, discrete; blade lanceolate, to 8 x 2.5 cm, acute at both ends; midrib evident and running into apex, lateral veins many and inconspicuous. Inflorescence a raceme to 3 cm long, with 3 or 4 pairs of triads and somewhat compressed axis; peduncle 1-1.5 cm long, triad peduncle 0.3-0.7 cm; bract and bracteoles small, triangular, persistent. Flowers not seen. Fruit ellipsoid, 0.8 x 0.4 cm, orange or brownish at maturity, calyculus inconspicuous.

Distribution: Brazil; ca. 10 collections studied; probably Guyana and French Guiana (see note below) (GU: 1; FG: 1).

Specimens examined: Guyana: E. Berbice-Corentyne Region, along Canje R., from Ekwarum Cr. to Three Sisters, Pipoly *et al.* 11417 (LEA, US). French Guiana: Saül, La Fumée Mt., La Fumée NE, Mori *et al.* 18185 (LEA, NY).

Note: The two fruiting specimens cited above are placed together and identified as *S. concinnus* on a provisional basis. They do not completely correspond with the circumscription of the species found in Krause (1932) with respect to the supposedly lenticellate stem surface and sessile triads, features which Krause may have based on Brazilian material (see the diagnosis and illustrations in Eichler, 1868). No flowers are present on either specimen. Pipoly *et al.* 11417 differs from Mori *et al.* 18185 in having a minutely papillose stem surface, the former being glabrous.

2. **Struthanthus dichotrianthus** Eichler in Mart., Fl. Bras. 5(2): 75. 1868. Type: Venezuela, near Tovar, Fendler 1119 (lectotype BR, designated by Kuijt 1994: 197).

Delicate, sparsely branched plants; internodes terete or slightly angled, some with epicortical roots to 6(-9) cm long. Petiole slender, 0.2-0.5 cm long; blade thin, narrowly lanceolate, mostly to 3(rarely -9) x 1(-3) cm, apex and base acute. Inflorescence a slender raceme of 1(2) pairs of triads; peduncle 0.4-0.7 cm, triad peduncle spreading, 0.3-0.4 cm long. Flowers sessile; bract and bracteoles small, acute, persistent. Male flower bud clavate, 0.6 cm long; female flower bud narrowly ellipsoid, ca. 0.5 cm long. Fruit ellipsoid, 0.8 x 0.5 cm, dull orange.

Distribution: Venezuela and throughout the Guianas, northern Brazil; ca. 70+ collections studied, of which ca. 40 from the Guianas (GU: 30; SU: 3; FG: 2).

Selected specimens: Guyana: Rupununi Savanna, near Toroebaroe Cr., Jansen-Jacobs *et al.* 1103 (LEA, U); Margins of Upper Abary R., Maas *et al.* 5441 (U), 5442 (NY, U); Rupununi Distr., Dadanawa, Rupununi R., savanna near river, Maas *et al.* 3674 (NY, U); Rupununi Northern Savanna, Mountain View Hill, Goodland & Persaud 686 (NY); Demerara-Mahaica Region, 3-4 km S of Loo Cr. along Linden Hwy, 32 km S of Timehri, Gillespie *et al.* 788 (LEA, US); Essequibo-Ils-W Demerara Region, Naamryck Canal, 3.5 km SW of Parika, Gillespie 1003 (LEA, US). Suriname: along Coronie road, between mouth of Coppename R. and Coronie, Lanjouw & Lindeman 1435 (K); in swamp between Hamptoncourt and Henarpolder, Nickerie, Lanjouw &

Lindeman 3175 (K); Bank of Nickerie R. opposite Nieuw Nickerie, Lanjouw & Lindeman 3129 (K); Lely Mts., SW plateau covered by ferrobauxite, forest along road N from airstrip, Lindeman & Stoffers *et al.* 452 (K). French Guiana: Site du future barrage de Petit Saut au pk 45 de la route d'accés, Billiet & Jadin 4486 (BR); Mt. Mahury, Hijman & Weerdenburg 200 (K).

Note: An extremely variable species both in leaf morphology and in its inflorescence.

3. **Struthanthus gracilis** (Gleason) Steyerm. & Maguire, Mem. New York Bot. Gard. 17(1): 443. 1967. – *Phthirusa gracilis* Gleason, Bull. Torrey Bot. Club 58: 359, f. 4. 1931. Type: Venezuela, Mt. Duida, Crest of Ridge 25, 6300 ft., Gleason 522 (holotype NY).

Struthanthus mucronatus Steyerm. in Steyerm. *et al.*, Bol. Soc. Venez. Ci. Nat. 26: 418. 1966. – *Struthanthus gracilis* (Gleason) Steyerm. & Maguire var. *mucronatus* (Steyerm.) Rizzini in Luces & Steyerm., Fl. Venez. 4(2): 17. 1982. Type: Venezuela, Bolívar, Steyermark *et al.* 92336 (isotype US).
Struthanthus chimantensis Steyerm. & Maguire, Mem. New York Bot. Gard. 17(1): 443. 1967. Type: Venezuela, Bolívar, Steyermark 74816 (holotype VEN).
Struthanthus cupulifer Rizzini, Bol. Soc. Venez. Ci. Nat. 32: 328. 1976. Type: Venezuela, Bolívar, Steyermark *et al.* 109825 (holotype RB).

Sparsely branching, smooth plants; young twigs much elongated (up to 10 cm), bearing recurved leaves on prehensile petioles; internodes terete, often bearing epicortical roots. Petiole 0.5-0.7 cm long; mature blade rather coriaceous, lance-elliptic to obovate, to 6 x 3.5 cm, apex acute to obtuse or mucronate, base acute; venation pinnate. Inflorescence mostly solitary in leaf axils, to 6 cm long, bearing 5-7 pairs of sessile triads; bract and bracteoles short and blunt, transformed into small, shallow cupules. Flower 0.4 cm long, pale green; style nearly as long as petals, stigma well developed, capitate. Fruit ellipsoid, at least 0.5 x 0.35 cm, bright yellow when mature.

Distribution: Venezuela (Amazonas, Bolívar) and Guyana; 15 collections studied (GU: 12).

Selected specimens: Guyana: Cuyuni-Mazaruni Region, Imbaimadai, Partang R. crossing, 1 km from mouth, Gillespie *et al.* 2693 (LEA, US); Cuyuni-Mazaruni Region, Pakaraima Mts., 1.5-2 km ESE of Imbaimadai trail along Partang R., Hoffman *et al.* 1982 (LEA, US);

Cuyuni-Mazaruni Region, Waleliwatipu (Aymatoi), N of Paruima, Clarke 996 (LEA, US); Cuyuni-Mazaruni Region,Chi-Chi Mt. Range, 2 km W of Chi-Chi Falls, S bank of Mazaruni R., Pipoly *et al.* 10264 (LEA, NY, US), 10273 (LEA, NY, US).

Note: This species is very reminiscent of a small *S. orbicularis* (Kunth) Blume, to which it is undoubtedly closely related; especially the young twigs look very much like slender ones of *S. orbicularis*. It differs in its smaller leaves and curious, sessile triads, the fruiting pedicels not accrescing.

4. **Struthanthus syringifolius** (Mart.) Mart., Flora 13: 102. 1830, as 'syringaefolius'. – *Loranthus syringifolius* Mart. in Schult. & Schult. f., Syst. Veg. 7: 141. 1829, as 'syringaefolius'. Type: Brazil, Amazonas, Martius s.n. (holotype M). – Fig. 22

Struthanthus heterophyllus Rizzini, Revista Brasil. Biol. 10: 42. 1950. Type: Guyana, Kalakoon, Jenman 2531 (holotype BM?, isotype K).

Rather coarse, sparsely branched plants, straight stems more or less terete at least when mature. Petiole to 2 cm long, distinct; blade mostly broadly lanceolate, to 15 x 7 cm, apex acute to somewhat caudate, base acute to slightly obtuse; venation evident, pinnate. Inflorescence usually 1 per leaf axil, to 6 cm long, nearly half of which may be peduncle; axis somewhat angular, bearing 3-6(8) pairs of triads; triad peduncle perpendicular, to 0.5(-1) cm long, terminating in a persistent involucre of 1 bract and 2 bracteoles each 0.15-0.3 cm long, in the female accrescing to about twice that size. All flowers sessile; mature flower buds 0.7-0.9 cm long; ovary 0.1-0.2 cm long, style straight or slightly undulate in upper half, 0.3-0.6 cm long, stigma capitate. Fruit ovoid, to 1.2 x 0.8 cm.

Distribution: Eastern Venezuela, the Guianas, Brazil; ca. 40 collections studied, ca. 30 of which are from the Guianas (GU: 12; SU: 20; FG: 1).

Selected specimens: Guyana: Bartica-Potaro road, 83 mi, Tutin 205 (K, U); Corentyne R., Acalla, Im Thurn s.n. (K); Kalacoon, Jenman 2531 (K); Demerara R., Jenman 5109 (K); Basin of Rupununi R., A.C. Smith 2435 (K); Rupununi Savanna, N of Kanuku Mts., dry savanna NE of Lethem airstrip, Lanjouw *et al.* 930 (U); Pakaraima Mts., Kamarang, Maas *et al.* 3953 (U). Suriname: surroundings of Uitkijk, 20 km from Paramaribo, den Outer 879 (U); between mouth of Coppename R. and

Fig. 22. *Struthanthus syringifolius* (Mart.) Mart.: fruiting habit (based on: Anderson 7252).

Coronie, km 3.1, Lanjouw & Lindeman 1494 (U); Plantage Dordrecht, lower Suriname R., A. Mennega 162 (U); near Paramaribo, Focke 644 (U); Lower Suriname R. near Plantation Marienburg, Went 269 (U). French Guiana: Cayenne, Martin s.n. (K).

8. **TRIPODANTHUS** (Eichler) Tiegh., Bull. Soc. Bot. France 42: 178. 1895. – *Phrygilanthus* subg. *Tripodanthus* Eichler in Mart., Fl. Bras. 5(2): 48. 1868.
   Type: T. acutifolius (Ruiz & Pav.) Tiegh. (Loranthus acutifolius Ruiz & Pav.)

Large, glabrous plants parasitic on branches and roots of numerous trees, with epicortical roots produced from the base and from some older stems. Leaves decussate, petiolate. Inflorescences axillary and/or terminal racemes of triads. All flowers pedicellate, perfect, 6-merous, stamens dimorphic, epipetalous, with versatile, dorsifixed anthers. Fruit an endospermous berry. (x = 8).

Distribution: A genus of 3 species in S America: *T. acutifolius* with a wide distribution from Venezuela to Bolivia, apparently not including most of the Pacific slopes, to northern Argentina, *T. flagellaris* (Cham. & Schlecht.) Tiegh., restricted to north-central Argentina and adjacent Uruguay and Brazil and a third, red-flowered, Colombian species has recently been described by Roldán & Kuijt (Novon 15: 207-209. 2005).

1. **Tripodanthus acutifolius** (Ruiz & Pav.) Tiegh., Bull. Soc. Bot. France 42: 179. 1895. – *Loranthus acutifolius* Ruiz & Pav., Fl. Peruv. Chil. 3: 48, t. 274, f. b. 1802. – *Phrygilanthus acutifolius* (Ruiz & Pav.) Eichler in Mart., Fl. Bras. 5(2): 49. 1868. Type: not designated.
   – Fig. 23

Woody liana on branches and trunks of trees and shrubs, sometimes reaching roots of host and becoming many meters long; epicortical roots numerous and prominent at base of plant and on some older parts of stem. Leaf blade thin to somewhat leathery, lanceolate to ovate, 4.5-10 x 1-3.5 cm, variably acuminate. Inflorescence usually 1 per leaf axil but also terminal, an (often) determinate raceme of paired, pedunculate triads, pedicels in a triad of equal length; bracts and bracteoles minute, caducous. Flowers bisexual, 6-merous; petals 1-1.2 cm long; filaments attached to petals slightly below middle, anthers short, versatile, dorsifixed; ovary narrow, calyculus inconspicuous, style straight, stigma undifferentiated. Fruit blackish, ovoid, about 0.7 cm long; endosperm copious, white; embryo slender, dicotylous, haustorial disk scarcely differentiated at maturity.

Distribution: In northern S America rarely collected, only from a few widely spaced areas, becoming more common further south in Bolivia, S Brazil and northern Argentina; no authentic collections have been seen from the Guianas, but the species may be expected at least at the higher

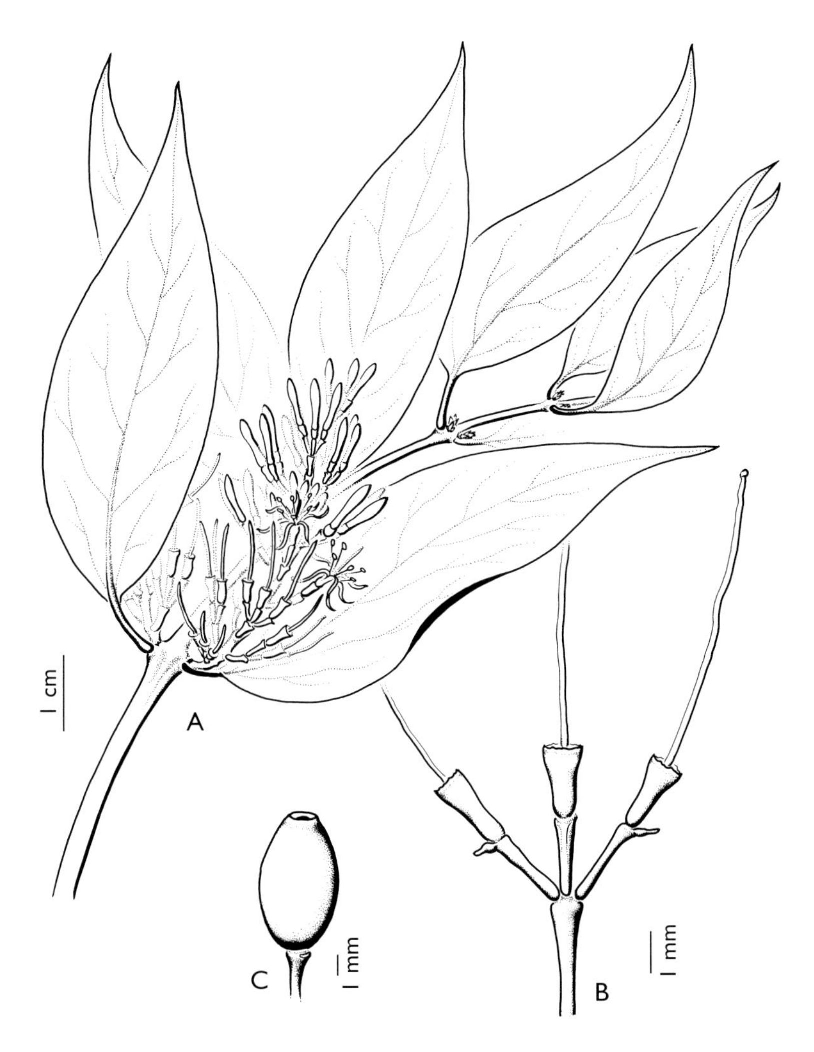

Fig. 23. *Tripodanthus acutifolius* (Ruiz & Pav.) Tiegh.: A, habit; B, inflorescence triad shortly after flowering; C, fruit (reproduced from Kuijt, Fl. of Ecuador 24, 1986).

levels in the Pakaraima Mts. area, since it occurs just west of Mt. Roraima in Venezuela, at 2000 m elev. (Huber & Steyermark 6939, LEA).

Note: The plant can become very large and, like *Gaiadendron*, can be seemingly terrestrial even though, in contrast to the latter, this habit develops following establishment on an aerial branch of a woody host, the epicortical roots then growing downwards into the soil.

106. VISCACEAE

by

Job Kuijt

Leafy to squamate plants. Lacking epicortical roots, implanted by means of primary haustoria only, these often expanding by means of endophytic strands. Stems often (but inconspicuously) articulated at nodes, brittle. Leaves decussate. Monoecious or dioecious. Flowers sunken in inflorescence axis, unisexual, 2-4-merous, at least female without aborted organs of opposite sex; male flower with perianth segments around small central disk or prominence; anthers isomorphic, sessile, with 1, 2, or numerous locules. Fruit a 1-seeded berry. (x = 10, 11, 12, 13, 14).

Distribution: A rather large family of 7 genera, tropical and/or temperate, all genera except *Arceuthobium* (restricted to the Northern Hemisphere) being limited to either the Old or the New World.; 34 species in 2 genera in the Guianas.

LITERATURE

Barlow B.A. 1964. Classification of the Loranthaceae and Viscaceae. Proc. Linn. Soc. New South Wales ser. 2. 89: 268-272.

Dixit, S.N. 1962. Rank of the subfamilies Loranthoideae and Viscoideae. Bull. Bot. Surv. India 4: 49-55.

Eichler, A.W. 1868. Loranthaceae. In C.F.P. von Martius, Flora Brasiliensis 5(2): 1-135.

Engler, H.G.A. & K. Krause. 1935. Loranthaceae. In H.G.A. Engler & K.A.E. Prantl, Die natürlichen Pflanzenfamilien, ed. 2. 16b: 98-203.

Krause, K. 1932. Loranthaceae. In A.A. Pulle, Flora of Suriname 1(1): 4-24.

Kuijt, J. 1959. A study of heterophylly and inflorescence structure in Dendrophthora and Phoradendron (Loranthaceae). Acta Bot. Neerl. 8: 506-546.

Kuijt, J. 1964. A revision of the Loranthaceae of Costa Rica. Bot. Jahrb. Syst. 83: 250-326.

Kuijt, J. 1968. Mutual affinities of Santalalean families. Brittonia 20: 136-147.

Kuijt, J. 1978. Commentary on the mistletoes of Panama. Ann. Missouri Bot. Gard. 65: 736-763.

Kuijt, J. 1986. Viscaceae. In G. Harling & B. Sparre, Flora of Ecuador 24: 11-112.

Kuijt, J. 1994. Typification of the names of New World mistletoe taxa (Loranthaceae and Viscaceae) described by Martius and Eichler. Taxon 43: 187-199.
Kuijt, J. 1996. Cataphylls and taxonomy in Phoradendron and Dendrophthora (Viscaceae). Acta Bot. Neerl. 45: 263-277.
Kuijt, J. & E. A. Kellogg. 1996. Miscellaneous mistletoe notes, 20-36. Novon 6: 33-53.
Kuijt, J. 2005. Viscaceae. In J.A. Steyermark *et al.*, Flora of the Venezuelan Guayana 9: 464-492.
Lemée, A.M.V. 1955. Loranthacées. Flore de la Guyane Française 1: 535-544.
Lindeman, J.C. & A.R.A. Görts-van Rijn. 1968. Loranthaceae. In A.A. Pulle & J. Lanjouw, Flora of Suriname, Additions and corrections 1(2): 295-300.
Rizzini, C.T. 1956. Pars specialis prodromi monographiae Loranthacearum Brasiliae terrarumque finitimarum. Rodriguésia 18-19(30-31): 87-234.
Rizzini, C.T. 1978. Los generos venezolanos y brasileros de las Lorantaceas. Rodriguésia 30(46): 27-31.
Wiens, D. & B.A. Barlow. 1971. The cytogeography and relationships of the Viscaceous and Eremolepidaceous mistletoes. Taxon 20: 313-332.

KEY TO THE GENERA

1 Anthers 1-locular; lateral vegetative branches with or without basal cataphylls; flowers usually in 1 or 3 series above each fertile bract; if squamate, stems terete . . . . *1. Dendrophthora*
Anthers 2-locular; lateral vegetative branches nearly always with basal cataphylls; flowers in 2 or more series above each fertile bract; if squamate, stems flattened or conspicuously keeled . . . . *2. Phoradendron*

1. **DENDROPHTHORA** Eichler in Mart., Fl. Bras. 5(2): 102. 1868.
Type: D. opuntioides (L.) Eichler (Viscum opuntioides L.)

Usually small, glabrous bushes, parasitic on dicotyledonous woody plants; internodes terete or somewhat angled; haustorial attachments at least initially simple, but cortical strands reaching into host tissues to various distances. Leaves simple, often rather succulent, in some species reduced to scales; lowest leafy organs on lateral branches and spikes in median or transverse plane. Vegetative lateral branches often, but not always, with basal cataphylls, but inflorescences less commonly so. Inflorescences axillary (very rarely terminal), in axils of leaf scales or foliage leaves, solitary or in clusters, rarely in compound inflorescences, being a spike with a peduncle of at least one internode and one or more terminal, fertile internodes with flowers produced, as in *Phoradendron*, in intercalary fashion; flower arrangement various but commonly in 1 or 3 series above each fertile bract. Monoecious or dioecious, in the former case with diverse patterns of sex distribution on spikes and/or individuals. Anthers presumedly 1-locular, dehiscing by a transverse slit. Fruit commonly white, with persistent perianth segments; endosperm copious, bright green, as is the weakly differentiated, dicotylous embryo. (x = 14).

Distribution: A neotropical genus of some 120 species, ranging from southern Mexico and much of the Caribbean to Peru and Bolivia; on the continent it is characterized by a rather strict preference for high elevations; in the Guianas 7 species.

Notes: *Dendrophthora* is very closely allied to the much larger genus *Phoradendron* (see comments under the latter), and a number of species have been transferred from one to the other. The only character which distinguishes the two at present is the 1-locular vs. 2-locular anther, respectively, a character which, notwithstanding its practical problems, seems to be rather consistently reliable: i.e., it at least separates two large, natural assemblages, even though in some species of *Dendrophthora* not clearly belonging to the "typical" group the validity of this solitary character needs to be tested.
Dendrophthora and *Phoradendron* do not overlap ecologically nearly as much as their geography would suggest, *Phoradendron* remaining at low to middle elevations. *D. elliptica* and *D. warmingii* would seem to constitute exceptions to this altitudinal pattern of separation.

LITERATURE

Kuijt, J. 1961. A revision of Dendrophthora (Loranthaceae). Wentia 6: 1-145.

Kuijt, J. 1963. Dendrophthora: additions and changes. Acta Bot. Neerl. 12: 521-524.

Kuijt, J. 2000. An update on the genus Dendrophthora (Viscaceae). Bot. Jahrb. Syst. 122: 169-193.

KEY TO THE SPECIES

1 Leaves < 2.5 x 1 cm; inflorescences < 1 cm long, peduncle simple 2
Leaves > 2 x 1 cm; inflorescences > 2 cm long, peduncle with 1-3 pairs of basal cataphylls 4

2 Flowers 1-seriate . . . . *6. D. roraimae*
Flowers 2- or 3-seriate 3

3 Leaves linear to very narrowly lanceolate, with very acute apex; fruit ovoid, smooth when young . . . . *1. D. decipiens*
Leaves broadly obovate to elliptical, apex rounded; fruit globose, often warty when young or fresh . . . . *3. D. elliptica*

4 Flowers 2-seriate or mostly so 5
Flowers 3-seriate 6

5 Plants yellowish, percurrent, without intercalary cataphylls; inflorescences axillary only; leaves with rather obscure, palmate venation; Brazil nearly exclusively . . . . *7. D. warmingii*
Plants green, mostly dichotomous, the few percurrent shoots with intercalary cataphylls; terminal inflorescences present; venation completely obscure; known from Guyana only . . . . *5. D. perfurcata*

6 Flowers yellow, contrasting with inflorescence axis; inflorescence peduncle ca. 1 cm long; percurrent shoots with 1-3 pairs of intercalary cataphylls . . . . *4. D. fanshawei*
Flowers of same greenish color as inflorescence axis; inflorescence peduncle < 0.5 cm long; intercalary cataphylls absent . . *2. D. densifrons*

1. **Dendrophthora decipiens** Kuijt, Proc. Kon. Ned. Akad. Wetensch., Biol. Chem. Geol. Phys. Med. Sci. 93: 131. 1990. Type: Venezuela, Steyermark 105158 (holotype VEN).

Diffusely branched plants, yellowish to deep green; internodes terete, to 3 cm long but mostly much shorter; basal cataphylls 1 pair, extremely low (sometimes to 0.3 cm above axil on larger secondary branches) and inconspicuous. Petiole indistinct and very short; blade linear to narrowly lanceolate, to 2 x 0.2 cm, acute. Monoecious, male and female flowers on same inflorescence, former often above latter. Inflorescence mostly solitary, to 0.4 cm long; peduncle 0.05-0.1 cm; fertile internodes 1 or 2; flowers 1-3 per fertile bract, 2- or 3-seriate, rarely 1-seriate. Fruit white, ovoid, 0.25 x 0.2 cm, perianth segments erect.

Distribution: Tepui summits of Venezuela (Amazonas & Bolívar); Venezuela/Guyana border area, summit of Mt. Roraima; not yet recorded for the Guianas.

Note: Similar to *D. roraimae* but differing in its larger size and its smaller inflorescences with extremely short peduncles and mostly 2- or 3-seriate flowers.

2. **Dendrophthora densifrons** (Ule) Kuijt, Novon 6: 33. 1996. – *Phoradendron densifrons* Ule in Pilg., Notizbl. Bot. Gart. Berlin-Dahlem 6: 292. 1915. Type: Venezuela/Guyana, Mt. Roraima, 1900 m, Ule 8602 (holotype? MG).

Plants dark brown when dry, densely leafy; basal cataphylls 1 or 2 pairs. Petiole 0.4-0.7 cm long; blade ovate-elliptical, 3-6.5 cm long, about half as wide. Monoecious. Inflorescence to 2.5 cm long, with 1 pair of sterile cataphylls and 3 fertile internodes; flowers 4-6 per fertile bract, 3-seriate. Fruit unknown.

Distribution: Known from only 3 collections, 2 from Venezuela (Bolívar, Chimantá Massif), the other the type (Mt. Roraima); probably present in adjacent Guyana.

3. **Dendrophthora elliptica** (Gardner) Krug & Urb. in Urb., Ber. Deutsch. Bot. Ges. 14: 285. 1896. – *Viscum ellipticum* Gardner, London J. Bot. 4: 106. 1845. – *Phoradendron ellipticum* (Gardner) Eichler in Mart., Fl. Bras. 5(2): 119. 1868. Type: Brazil, Gardner 437 (holotype K, isotype G). – Fig. 24

Dendrophthora rubicunda Ule in Pilg., Notizbl. Bot. Gart. Berlin-Dahlem 6: 289. 1915. Type: Venezuela/Guyana, Mt. Roraima, Ule 8601 (holotype B, destroyed, isotype K).
Phoradendron dendrophthora Rizzini, Revista Fac. Agron. (Maracay) 8: 87. 1975. Type: Venezuela, Bolívar, Steyermark & Dunsterville 104351 (holotype RB).

Profusely branched, glabrous, small-leaved plant; basal cataphylls 1 pair, to 0.5 cm above axil, in median position. Leaf blade elliptical to somewhat ovate, to 2 x 1 cm. Monoecious. Inflorescence ca. 0.3 cm long; peduncle 0.1 cm, simple; fertile internode usually 1, normally with 5 flowers per fertile bract, 3-seriate, lowest several female. Fruit white, ovoid to nearly spherical, 0.5 x 0.6 cm, tuberculate when young, perianth segments small, erect.

Distribution: Throughout montane (mostly western) S America, to Bolivia, sometimes descending to middle elevations; ca. 70 collections studied, of which at least 3 from Guyana (GU: 3).

Specimens examined: Guyana: Pakaraima Mts., Mt. Aymotoi, Maas *et al.* 5784 (U, VEN); Mt. Roraima, path to upper savanna, Im Thurn 276 (US); Roraima, 1300 m, Ule 8601 (K).

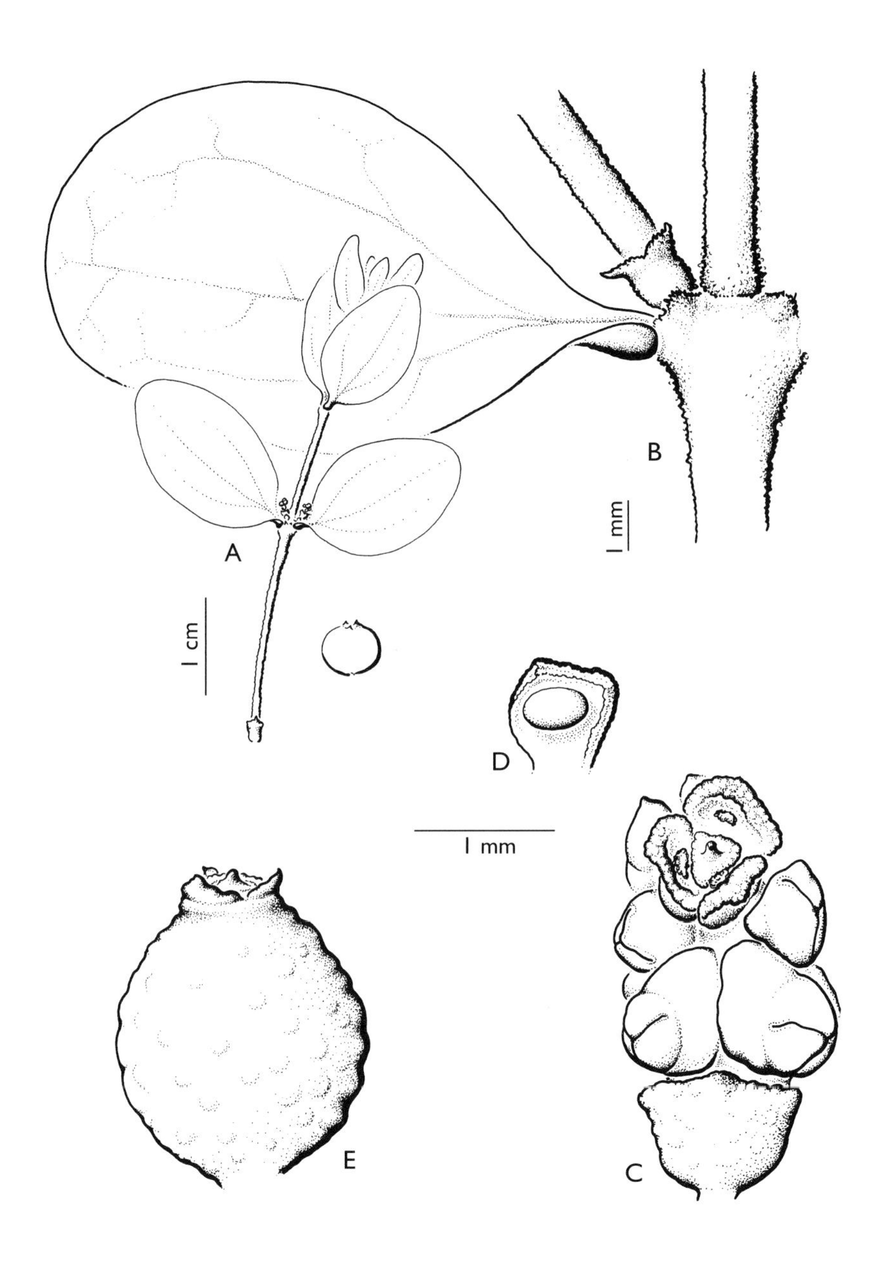

Fig. 24. *Dendrophthora elliptica* (Gardner) Krug & Urb.: A, habit and mature fruit; B, detail of basal cataphylls and leaf; C, inflorescence; D, petal and anther; E, immature fruit (reproduced from Kuijt, Fl. of Ecuador 24, 1986).

4. **Dendrophthora fanshawei** (Maguire) Kuijt, Novon 6: 33. 1996. – *Phoradendron fanshawei* Maguire in Maguire *et al.*, Bull. Torrey Bot. Club 75: 300. 1948. Type: Guyana, Maguire *et al.* 27201 (holotype NY, isotype K).

Stout plants, dichotomous and/or percurrent?; each innovation with 1 pair of foliage leaves and 2 or 3 pairs of cataphylls, highest ones reaching halfway to expanded leaves above; stems terete. Petiole 0.2-0.5 cm long; blade leathery, obovate, to 8 x 3.5 cm, apex rounded, base cuneately grading into massive, indistinct petiole; venation obscure. Dioecious. Female inflorescence 4 cm long; sterile internodes 1-3, fertile internodes 4; flowers 5-9 per fertile bract, 3-seriate.

Distribution: An infrequent species; 4 collections studied (GU: 1; SU: 3).

Selected specimens: Suriname: River margin, riverine forest, Lucie R., 2-5 km below confluence of Oost R., Irwin *et al.* 55486A (NY, U); Savanne, 2 km van Boven-Lucie R., langs lijn N to Wilhelmina Gebergte, Schulz 10369a (U).

Note: In the protologue of this species, branching is said to be dichotomous, but the specimens I have seen appear to be normally percurrent.

5. **Dendrophthora perfurcata** (Rizzini) Kuijt, Bot. Jahrb. Syst. 122: 178. 2000. – *Phoradendron perfurcatum* Rizzini in Maguire *et. al.*, Mem. New York Bot. Gard. 29: 34. 1978. Type: Guyana, Tillett *et al.* 45087 (holotype RB, isotypes MO, NY).

Plants mostly dichotomous; percurrent portions with 2 or 3 pairs of intercalary cataphylls, upper ones placed more or less halfway to next foliar node; basal cataphylls usually 2 pairs, upper pair somewhat below middle of innovation. Leaf blade obovate to elliptic, apex rounded; venation completely obscure. Inflorescences both terminal and axillary, to 2 cm long; 3 or 4 fertile internodes; flowers 3-7 per fertile bract, probably only 2-seriate. Fruit unknown.

Distribution: Known only from the type (GU: 1).

Note: The protologue states this species to have 3-seriate flowers, but my observations of the type indicate that they are 2-seriate, instead.

6. **Dendrophthora roraimae** (Oliv.) Ule, Bot. Jahrb. Syst. 52, Beibl. 115: 49. 1914. – *Phoradendron roraimae* Oliv. in Thurn, Timehri 5: 201. 1886. Type: Guyana, edge and summit of Mt. Roraima, Im Thurn 323 (holotype K, isotype US).

Small, erect plants; basal cataphylls 1 pair, to 1 cm above axil. Leaf blade rather thick, linear to narrowly lanceolate, to 2 cm long. Inflorescence about half as long as subtending leaf, axillary only; peduncle 0.2 cm long; fertile internode 1(2); flowers 3 or 4 per fertile bract, 1-seriate. Fruit white, spherical, perianth segments reflexed.

Distribution: Tepuis of Venezuela (Amazonas & Bolívar), Mt. Roraima; 8 specimens studied (GU: 4).

Selected specimens: Guyana: the type; Mt. Roraima, 8600 ft, McConnell & Quelch 82, 680 (K); Mt. Roraima, Ule 8599 (K).

7. **Dendrophthora warmingii** (Eichler) Kuijt, Novon 13: 88. 2003. – *Phoradendron warmingii* Eichler, Vidensk. Meddel. Dansk Naturhist. Foren. Kjøbenhavn 1870: 209. 1870. Type: Brazil, Lagoa Santa, Warming 383 (holotype C). – Fig. 25

Phoradendron tepuianum Steyerm. in Steyerm. *et al.*, Fieldiana Bot. 28: 223. 1951. – *Dendrophthora tepuiana* (Steyerm.) Kuijt, Proc. Kon. Ned. Akad. Wetensch., Biol. Chem. Geol. Phys. Med. Sci. 93: 138. 1990. Type: Venezuela, Bolívar, Steyermark 59688 (holotype F, isotype US).

Olive green to tawny or bronze plants, in part covered with innumerable yellow spots giving a granular appearance; young internodes angled or compressed-keeled soon becoming terete, often stout, to 6 cm long; nodes slightly swollen; branches with 1 pair of basal cataphylls 0.6-1 cm above axil, rarely with a second pair. Petiole ca. 0.5 cm long; blade coriaceous, ellipsoid to ovate, to 9 x 7 cm, apex obtuse to rounded, base tapering into stout, indistinct petiole; venation obscure but apparently basal-palmate. Inflorescence fairly stout, to 4 cm long; 1 or 2 pairs of basal cataphylls about 0.2 cm apart, followed by 3 or 4 fertile internodes, distal one mostly very short and rounded apically, larger ones with 5-9 flowers per fertile bract, usually 2-seriate. Fruit broadly ovoid, 0.4 x 0.4 cm, perianth segments more or less open.

Distribution: Brazil, Amazonian Peru, Venezuela, the Guianas; 25 specimens studied, 3 from the Guianas (GU: 2; FG: 1).

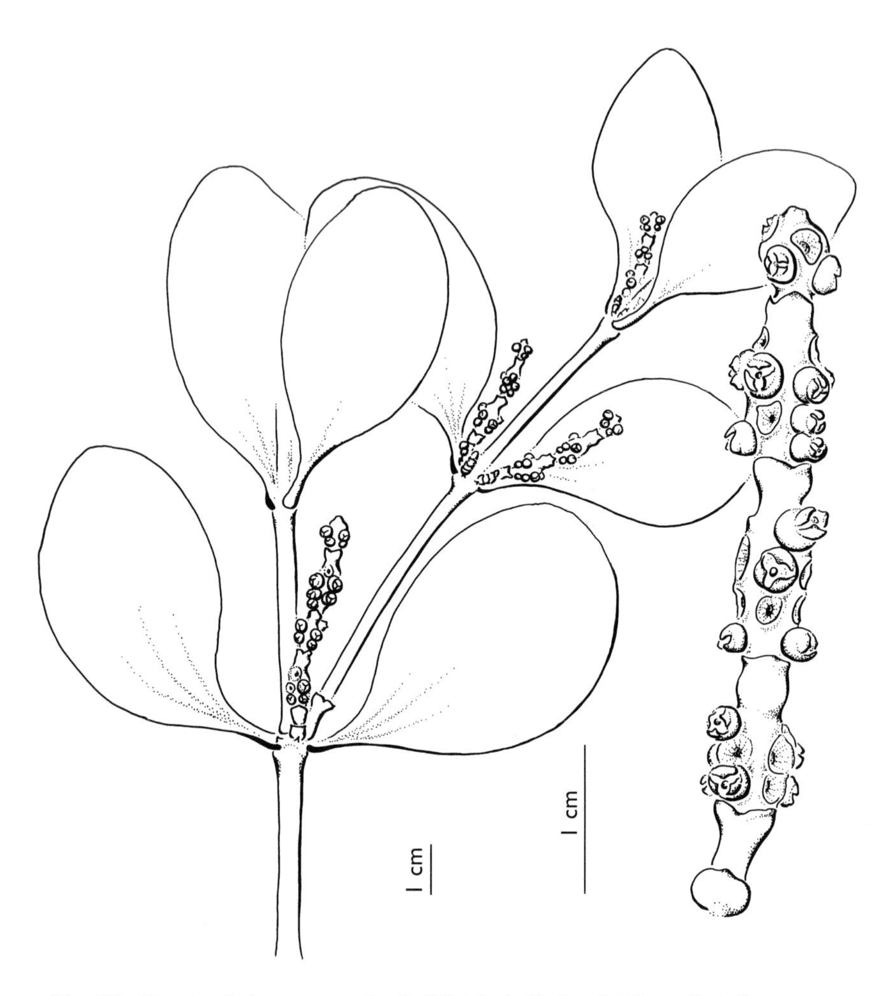

Fig. 25. *Dendrophthora warmingii* (Eichler) Kuijt: habit and inflorescence (reproduced from Kuijt, Bot. Jahrb. Syst. 122, 2000).

S p e c i m e n s e x a m i n e d: Guyana: Cuyuni-Mazaruni Region, foothills immediately S of Mt. Ayanganna, 1 km W of Pong Cr., Pipoly *et al.* 10693 (LEA); Paruima, 17 km W, 3 km W of eastern summit of Waukauyeng-tipu, Clarke *et al.* 5776 (LEA). French Guiana: Commune de Montsinnery, piste de Risque tout à 12 km de la route du Tour de l'Ile, Cremers 6020 (CAY, LEA).

2. **PHORADENDRON** Nutt., J. Acad. Nat. Sci. Philadelphia, ser. 2. 1: 185. 1848.
Type: P. californicum Nutt.

Leafy parasitic shrubs; nodes often with inconspicuous constriction; internodes terete or angled. Leaves entire, petiolate or sessile, venation pinnate to palmate. Base of lateral vegetative branches with one or more pairs of fused, basal cataphylls, some species with one or more pairs of intercalary cataphylls separating successive pairs of foliage leaves; cataphylls in most species sterile; branching mostly percurrent, sometimes dichotomous by means of terminal inflorescences or abortion of apex. Inflorescences spike-like; peduncle simple or with several squamate, sterile internodes, followed by one to several fertile internodes; each fertile bract with 2 or 3 longitudinal series of flowers, produced basipetally by intercalary action, number of flowers variable but flower area always topped by a single median flower. Flowers small, sessile and often sunken in small axial cavities, 3-(4-)merous, occurring in diverse monoecious or dioecious patterns. Anthers sessile, 2-locular, minute. Fruit a berry, smooth to warty, mostly white, yellowish or red, perianth segments of female flower persistent, closed, erect, or spreading; seed with copious bright green endosperm and small, dicotylous embryo. (x = 14).

Distribution: A neotropical genus with at least 234 species; the geographical range is very large, reaching from southern U.S.A, Mexico, through C America and the Caribbean to Peru, Bolivia, and north-central Argentina; in the Guianas 27 species.

Notes: *Phoradendron* is an extremely challenging genus, partly because of innumerable uncritical species descriptions in the past, and partly because of inherent complexities. The genus is often decidedly difficult to separate from the closely related genus *Dendrophthora,* a separation based largely on the latter's 1-locular anther vs. 2-locular in *Phoradendron.* The anther is extremely small and this criterion makes it technically impossible to place novelties which lack male flowers. A second difficulty is the large number of species of *Phoradendron.*
Flower seriation is especially important in the genus, and is usually quite constant. However, there are cases where a great deal of caution is needed in interpretation. This is especially so in those cases where some or all fertile internodes have only 3 flowers per fertile bract (1 apical flower and 2 representing lateral series). This pattern is found in *P. mucronatum*, for example, and there is no way to tell whether, if more flowers had been present, the median series would be represented or not (in other words, whether the inflorescence is 3-seriate or 2-seriate, respectively). In other

cases, as in *P. obtusissimum*, we may encounter some fertile internodes with both 3 (or even 1) and more than 3 flowers per bract; in such cases, we are able to discover the seriation type by inspecting the largest fertile internodes. Occasionally there is a great deal of sexual dimorphism, complicating the construction of keys. The key which follows is based on ascertaining the seriation type wherever possible.

LITERATURE

Kellogg, E.A. & R.A. Howard. 1986. The West Indian species of Phoradendron (Viscaceae). J. Arnold Arbor. 67: 65-107.

Kuijt, J. 2003. A monograph of Phoradendron. Syst. Bot. Monogr. 66: 1-643.

Rizzini, C.T. 1978. El género Phoradendron en Venezuela. Rodriguésia 30(46): 33-125.

Trelease, W. 1916. The genus Phoradendron.

KEY TO THE SPECIES

1 Leaves < 0.8 cm long, often reduced to scale-leaves, or apparently absent 2
Leaves foliaceous, more than 1 cm long 3

2 Small, erect plants with terete internodes 2-3 cm long; basal and intercalary cataphylls absent; fruiting perianth segments closed . . . . . . *2. P. aphyllum*
Pendent plants with compressed internodes 3-9 cm long; basal and intercalary cataphylls 1 pair each, tubular; fruiting perianth segments erect . . . . *20. P. poeppigii*

3 Plants dichotomous, either through terminal inflorescences or through abortion of apex 4
Plants percurrent (not dichotomous) 8

4 Flowers 3 per fertile bract, the terminal male, lateral ones female; leaves obscurely 3-veined; percurrent portions with 1 pair of intercalary cataphylls . . . . *25. P. strongyloclados*
Not the above sex distribution; leaves not obscurely 3-veined; percurrent shoots normally absent 5

5 Leaf blade mostly lanceolate and acute at both ends . . . . *13. P. morsicatum*
Leaf blade obovate, the apex broadly rounded 6

6 Flowers 1-3 per fertile bract, the male usually few and on lowest fertile internode; fruiting perianth segments erect or parted . . . . *10. P. inaequidentatum*
Flowers 5-9 per fertile bract, the single male highest above bract; fruiting perianth segments closed 7

7 Leaf blade to 6 x 3.5 cm, more or less leathery, venation obscure . . . . *15. P. northropiae*
Leaf blade to 13 x 10 cm, thin, venation evident . . . . *24. P. racemosum*

8 Intercalary cataphylls present; terminal inflorescences lacking 9
Intercalary cataphylls absent 12

9 At least some intercalary cataphylls subtending inflorescences 10
Intercalary cataphylls present but not subtending inflorescences 11

10 Peduncle simple or very unequally double; fertile internodes mostly 4; male flowers below female ones on each fertile internode; fruiting perianth segments closed . . . . *11. P. krameri*
Peduncle with 2 or more short and crowded sterile internodes, female flowers below (apical) male one in each flower area; fertile internodes 5-9; fruiting perianth segments more or less erect . . . . *6. P. crassifolium*

11 Plants golden green; young internodes sharply keeled; leaves with 3 or 5 palmate veins; flowers 3-seriate . . . . *5. P. chrysocladon*
Plants not golden green; internodes terete; venation pinnate; flowers mostly 2-seriate . . . . *19. P. piperoides*

12 Flowers 3 per fertile bract, apical one male, lateral ones female; fruit tuberculate . . . . *14. P. mucronatum*
Not the above flower number; male flowers in various positions, or plants dioecious; fruits smooth 13

13 Leaves with 3(5) basal veins, round-tipped; internodes quadrangular; flowers 2-seriate 14
Leaves with more than 3 basal veins, or venation pinnate, apex rounded to acute; internodes and flower seriation various 15

14 Fruiting perianth segments closed when mature, fruit yellow . . . . *23. P. quadrangulare*
Fruiting perianth segments erect when mature, fruit reddish orange . . . . *26. P. trinervium*

15 Internodes keeled to quadrangular below 16
Internodes terete at least below 21

16 Leaf apex acute or at least tapering; internodes keeled but not usually quadrangular or winged 17
Leaf apex somewhat tapering or rounded but not ultimately acute; internodes keeled to quadrangular or even winged 19

17 Leaf blade broadest below middle; plants monoecious; most fertile internodes with 5 flowers per fertile bract . . . . *27. P. undulatum*
Leaf blade broadest at middle; plants dioecious; flowers 1 or 3 per fertile bract, at least in female 18

18 Female flowers 3 per fertile bract; fertile internodes 2 or 3; fruit placed on middle of fertile internode; fruiting perianth segments somewhat parted to erect; known from Suriname only . . . . *1. P. acuminatum*
Female flowers 1 per fertile bract; fertile internodes 5-8; fruit placed distally on fertile internode; fruiting perianth segments closed; known from French Guiana only . . . . *8. P. granvillei*

19 Fowers 2-seriate 20
Flowers 3-seriate or 3 per fertile bract 23

20 Leaves obovate, of varnished appearance; inflorescence peduncle with 2-4 (-10) sterile internodes . . . . *20. P. pteroneuron*
Leaves lanceolate, not of varnished appearance; inflorescence peduncle simple . . . . *4. P. bilineatum*

21 Perhaps exclusively hyperparasitic on other mistletoes; petiole often broad, flat, blade somewhat amplexicaul . . . . *7. P. dipterum*
Not hyperparasitic; petiole not flat, blade not amplexicaul 22

22 Young leaves translucent red; inflorescence with 3(4) fertile internodes; fruits no more than 3 per fertile internode, placed high on fertile internode . . . . *17. P. pellucidulum*
Young leaves neither thin nor translucent red; inflorescence with 6-10 or more fertile internodes; fruits mostly > 3 per fertile bract, spread along fertile internode . . . . *18. P. perrottetii*

23 Flowers 2-seriate, more than 3 per fertile bract 24
Flowers 3-seriate, no more than 3 per fertile bract 26

24 Leaves coriaceous, venation completely obscure except for midrib . . . . *22. P. pulleanum*
Leaves not coriaceous, with evident venation 25

25 Flowers and (especially) fruits sunken in cavities in inflorescence axis; plants usually blackening when dry . . . . *3. P. bathyoryctum*
Flowers or fruits not sunken in cavities; plants not blackening usually . . . . *12. P. mairaryense*

26 Fruits placed very high on fertile internodes, 3 per fertile bract; fruiting perianth segments large, flaring . . . . *16. P. obtusissimum*
Fruits distributed along fertile internodes, mostly > 3 per fertile bract (except female *P. pellucidulum*); fruiting perianth segments relatively small, closed 27

27 Venation lacking strong basal, lateral veins; lower midrib usually lightly scurfy . . . . *9. P. hexastichum*
Venation palmate or pinnate with strong, basal lateral veins; lower midrib smooth, not scurfy 28

28 Young leaves thin, translucent red; inflorescence with 3(4) fertile internodes; mostly dioecious; fruit no more than 3 per fertile bract; placed high on fertile internode . . . . *17. P. pellucidulum*
Young leaves neither thin nor translucent red; inflorescence 4-10 or more fertile internodes; monoecious or dioecious; fruit mostly > 3 per fertile bract, spread along fertile internode 29

29 Normally hyperparasitic on other mistletoes; dioecious; internodes tending to be quadrangular to 4-winged; fruit 0.2 x 0.2 cm . . . . *7. P. dipterum*
Not hyperparasitic; monoecious; internodes no more than slightly keeled above; fruit 0.4 x 0.3 cm . . . . *18. P. perrottetii*

1. **Phoradendron acuminatum** Kuijt, Syst. Bot. Monogr. 66: 65. 2003. Type: Suriname, Tafelberg, Hawkins 1893 (holotype LEA). – Fig. 26

Slender-stemmed plants; internodes sharply keeled, 6-10 cm; basal catapylls 1 pair, 0.2-0.3 cm above base. Leaf blade thin, lanceolate, to 11 x 3 cm, apex and base sharply acute; venation pinnate, midvein running into apex. Dioecious, male plants not seen. Female inflorescence slender, ca. 1.2 cm long; peduncle unequally double or even triple, 0.4 cm; fertile internodes 2(3); flowers 3 per fertile bract, 2- or 3-seriate. Fruit white, ellipsoid, 0.04 x 0.025 cm, perianth segments small, somewhat parted to erect.

Distribution: Known only from the type (SU: 1).

2. **Phoradendron aphyllum** Steyerm. in Steyerm. *et al.*, Bol. Soc. Venez. Ci. Nat. 26: 412. 1966. Type: Venezuela, Bolívar, Steyermark & Nilsson 473 (holotype VEN). – Fig. 27

Small, leafless plants, orange-brown; internodes terete, stout, 2-3 cm long; nodes somewhat swollen; basal cataphylls absent. Monoecious. Inflorescences clustered at nodes, to 1.5 cm long in fruit; peduncle mostly simple, 0.2-0.4 cm, sometimes with several sterile internodes, fertile internodes 1-several, thick; flowers mostly 3 per fertile bract, apical flower above fertile bract male at least at times, others female. Fruit ellipsoid, 0.6 x 0.4 cm, perianth segments closed.

Distribution: Venezuela and adjacent Guyana; 4 collections studied (GU: 2).

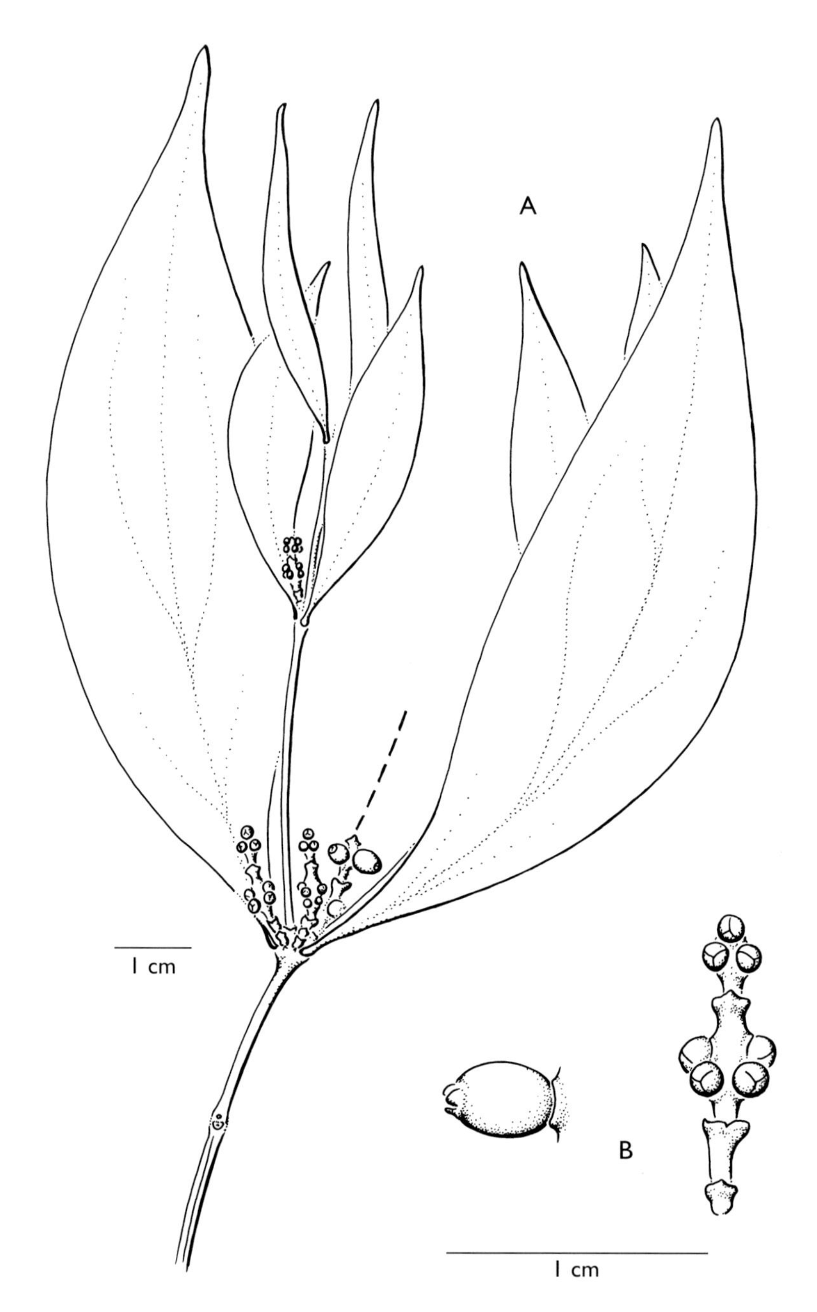

Fig. 26. *Phoradendron acuminatum* Kuijt: A, habit, the broken line represents a missing, percurrent shoot; B, infructescence and fruit (reproduced from Kuijt, Syst. Bot. Monogr. 66, 2003).

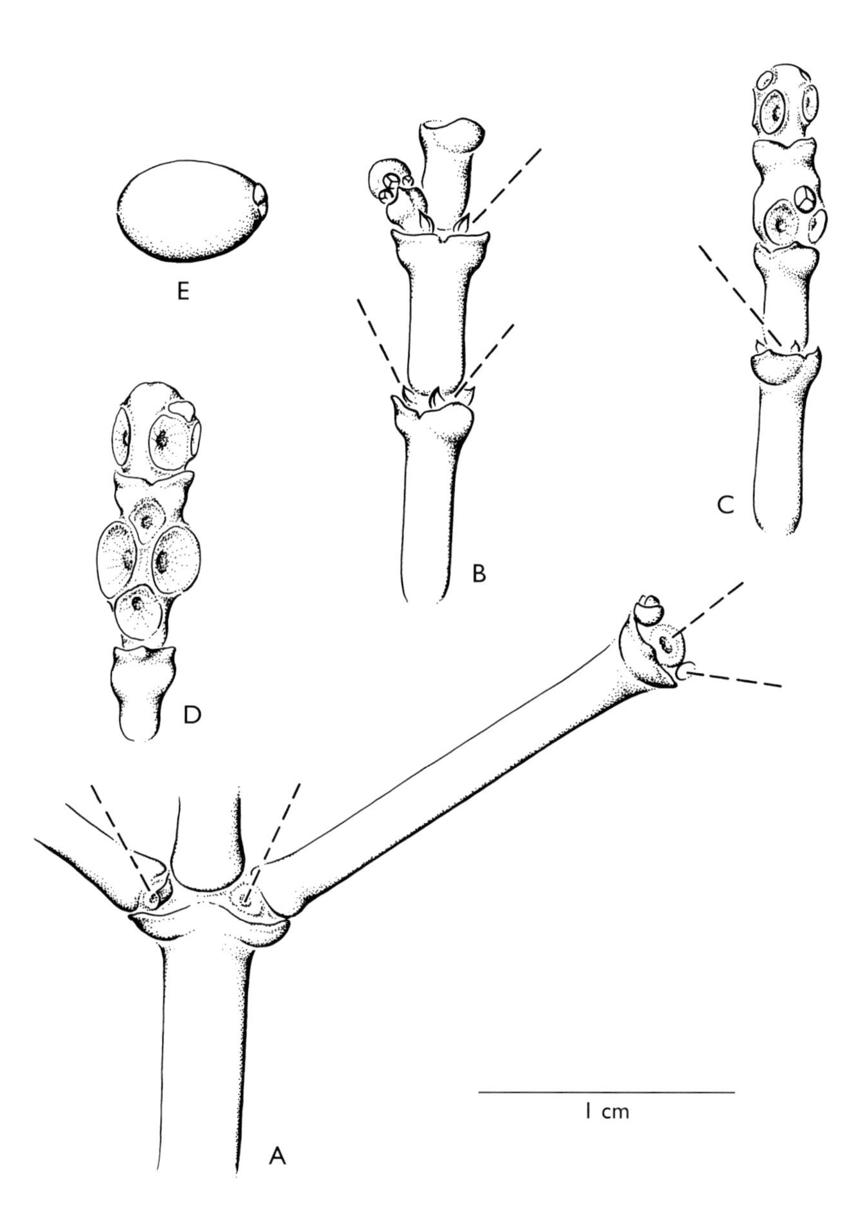

Fig. 27. *Phoradendron aphyllum* Steyerm.: A, part of shoot with lateral branch; B, terminal portion of shoot, possibly terminating in an inflorescence; C, terminal inflorescence; D, inflorescence; E, fruit (reproduced from Kuijt, Syst. Bot. Monogr. 66, 2003).

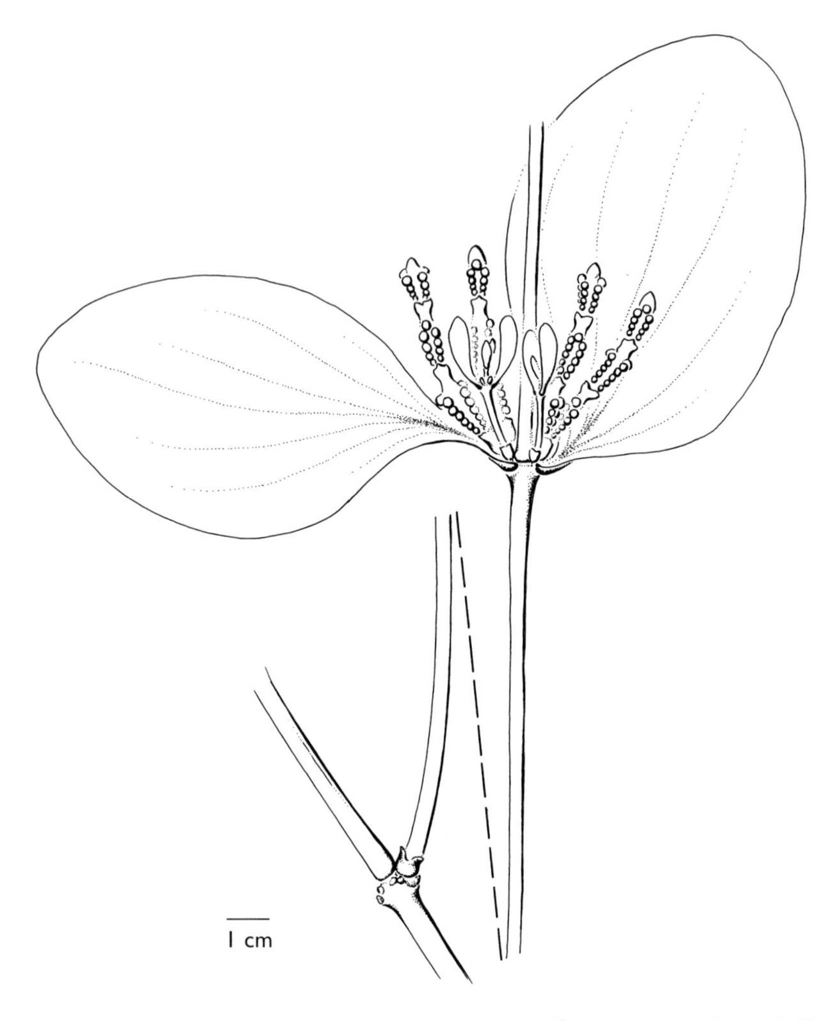

Fig. 28. *Phoradendron bathyoryctum* Eichler: habit (reproduced from Kuijt, Syst. Bot. Monogr. 66, 2003)

Specimens examined: Guyana: Upper Mazaruni R. basin, Merume Mts., Partang R., Tillett *et al.* 43964 (NY); Ridge behind Camp 3, 0.5 mi below "falls" of Kako R., S bank, Partang R. basin, Tillett *et al.* 45512 (NY, RB).

3. **Phoradendron bathyoryctum** Eichler in Mart., Fl. Bras. 5(2): 123. 1868. Type: Brazil, Piauí, Gardner 2626 (lectotype P, designated by Trelease 1916: 88). – Fig. 28

Rather large plants, percurrent; lacking intercalary cataphylls; internodes to 14(18) cm long, terete when mature, often sharply keeled to quadrangular when young; basal cataphylls 1 pair, ca. 0.5 cm above base. Petiole to 1 cm long; blade ovate to elliptical, to 15 x 7 cm, apex rounded, base contracted into indistinct petiole; venation more or less palmate. Monoecious, male flowers rare. Inflorescence to 5 cm long; peduncle 0.3-0.5 cm, simple (rarely double), followed by 3 or 4 fertile internodes with 9-13 flowers per bract in regular 2-seriate arrangement, partly immersed in axis; fused fertile bracts continuous with each other, showing scarcely a sinus. Fruit yellowish-orange, globose, ca. 0.3 x 0.3 cm, perianth segments closed.

Distribution: A very common species in Brazil, ranging into Paraguay and Amazonian Bolivia; ca. 200 collections studied (SU: 2).

Specimens examined: Suriname: along Paulus-kreek S of Paramaribo, Florschütz *et al.* 907 (U); near airport at Oelemari, along banks of Oelemari R., Wessels Boer 1120 (COL, K, U).

Vernacular name: Suriname: fowroedoti.

4. **Phoradendron bilineatum** Urb., Bot. Jahrb. Syst. 23, Beibl. 57: 5. 1897. Type: Venezuela, Aragua, Fendler 1811 (holotype B, destroyed, isotypes GH, GOET, K, MO). – Fig. 29

Phoradendron carinatum Trel., Phoradendron 139. 1916. Type: Guyana, Jenman 2542 (holotype K, isotype NY).

Internodes keeled, to 7.5 cm long; basal cataphylls 1 pair, ca. 0.2 cm above base, or with a second, slighter higher pair. Petiole 0.5-1 cm long; blade lanceolate, to 10 x 2.5 cm, apex more or less acute, base cuneately tapering to indistinct petiole. Infructescence to 2.5 cm long; peduncle simple, 0.2 cm long, or double and somewhat longer; fertile internodes 4 or 5; flowers 3-9 per fertile bract, 2-seriate. Fruit ellipsoid, ca. 0.3 x 0.25 cm, perianth segments closed.

Distribution: Venezuela, the Guianas, northern Brazil; 23 collections studied (GU: 9; SU: 3; FG: 2).

Selected specimens: Guyana: Upper Demerara, Mabura Hill, access to road Tropenbos Field Station, Christenson *et al.* 1875 (LEA); Potaro-Siparuni Region, Paramaktipu, Clarke *et al.* 1159 (LEA, US); MacKenzie, Cook 5 (K); Mabura Region, Pibiri Compartment, Plot 1, Ek 652 (LEA, U); Rupununi Northern Savanna, Moreru, Goodland 1066

Fig. 29. *Phoradendron bilineatum* Urb.: habit and young inflorescence (reproduced from Kuijt, Syst. Bot. Monogr. 66, 2003).

Fig. 30. *Phoradendron chrysocladon* A. Gray: habit (reproduced from Kuijt, Syst. Bot. Monogr. 66, 2003).

(LEA, NY, R). Suriname: Coppename R., in rapids above Sidomkroetoe, Geijskes s.n. (K); shores of Marowijne R., between Armina Falls and Lakama Falls, Lanjouw & Lindeman 3473 (K, U); along bank of Coppename R. near mouth of Wayombo R., Lindeman 6266 (MO, U). French Guiana: downstream from Roche Touatou, Oyapoque Basin, Cremers *et al.* 13933 (LEA, MO, U).

5. **Phoradendron chrysocladon** A. Gray, U.S. Explor. Exped., Bot. Phan. 743. 1854. Type: Brazil, Rio de Janeiro, Wilkes Exped. s.n. (holotype US, isotypes F, GH, P). – Fig. 30

Yellowish green plants; stems terete to slightly keeled; successive pairs of foliage leaves separated by 1 pair of very acute intercalary cataphylls to 1.5 cm above base; basal cataphylls of lateral branches 1 pair, acute, sometimes with a second pair. Petiole 0.5 cm long, indistinct, stout; blade broadly lanceolate, to 15 x 9 cm, apex acute; usually 5 conspicuous long basal veins. Monoecious, sex distribution variable. Inflorescence to 5 cm long, mostly with simple peduncle to 1 cm long; fertile internodes 3-5, flowers ca. 12 per fertile bract, usually very regularly 3-seriate. Fruit ovoid to spherical, 0.3 cm long, perianth segments closed.

Distribution: Variable species, known from the Caribbean, to Brazil and Amazonian Bolivia and Peru; 400+ collections studied (GU: 5).

Specimens examined: Guyana: Paruima, 15 km W, Clarke *et al.* 5761 (LEA); Paruima, savanna on Waukauyeng-tipu, Clarke *et al.* 5860 (LEA, US); Potaro-Siparuni Region, Pakaraima Mts., Mt. Wokomung, W face of ridge, headwaters of Wusupubaru Cr., 1 km downslope to Patamona Lookout, Henkel *et al.* 1403 (LEA); Pakaraima Mts., 12 m waterfall, large Partang R. tributary, 12.7 km NE of Imbaimadai, Hoffman *et al.* 1881 (LEA); Cuyuni-Mazaruni Region, Ayanganna Plateau, savanna N of montain center, Pipoly *et al.* 10889 (LEA, US).

6. **Phoradendron crassifolium** (Pohl ex DC.) Eichler in Mart., Fl. Brasil. 5(2): 125. 1868. – *Viscum crassifolium* Pohl ex DC., Prodr. 4: 280. 1830. Type: Brazil, Pohl 457 (holotype G-DC). – Fig. 31

A large and coarse plant with terete stems; lateral branches with 1 or 2 pairs of basal cataphylls to 0.5 cm above axil, percurrent branches with 1 pair of intercalary, sterile cataphylls > 0.5 cm above node, followed by 2-5 pairs of fertile cataphylls, these subtending 1-3 inflorescences each, latter also present in axils of foliage leaves. Petiole short, heavy; blade leathery, broadly lanceolate, to 17 x 10 cm; 3-7 prominent palmate veins. Monoecious. Spikes 2-3.5 cm long in fruit, with 2-6(-10) crowded pairs of sterile cataphylls at base; fertile internodes 5-9, very short and crowded, each fertile bract with 3-7 flowers in 2-seriate pattern, terminal flower male, lateral ones female. Fruit yellowish or pale orange to white, often reddish-brown where fully exposed to sun, smooth, nearly spherical, about 0.4 x 0.3 cm, perianth segments small, erect.

Distribution: C America and S America, south to Bolivia and Paraguay; usually at low elevations; 500+ collections studied (GU: 45; SU: 7; FG: 11).

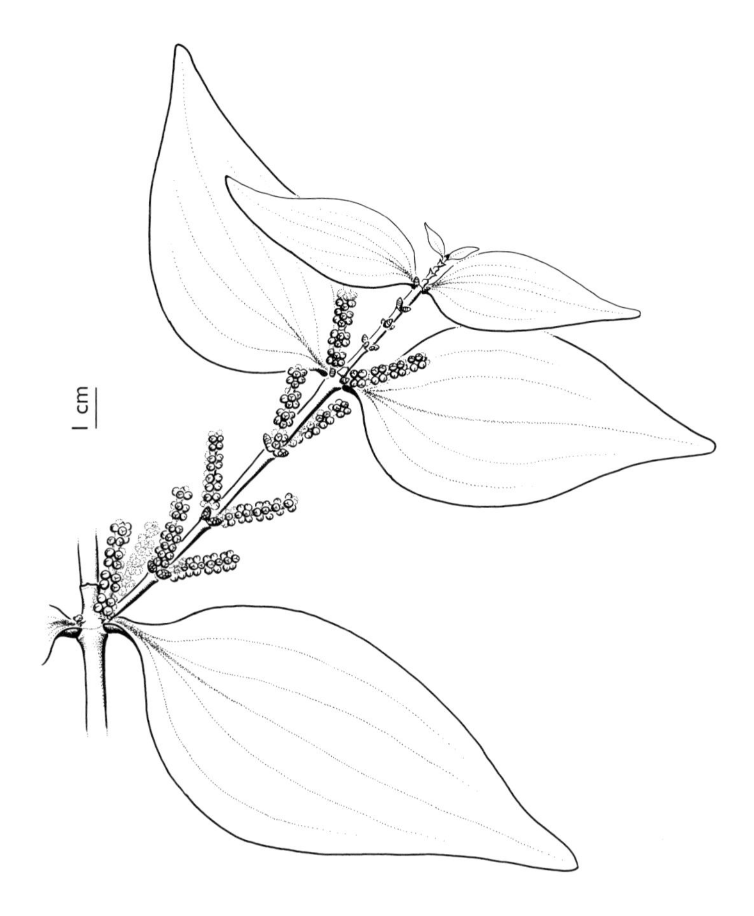

Fig. 31. *Phoradendron crassifolium* (Pohl ex DC.) Eichler: habit (reproduced from Kuijt, Syst. Bot. Monogr. 66, 2003).

Selected specimens: Guyana: Berbice Region, along banks of Corentyne R., above Baba Grant Sawmill, above Cow Falls, McDowell *et al.* 2336 (LEA); Upper Canje R., along S bank from abandoned camp W to Digitima Cr. mouth, Pipoly *et al.* 11520 (LEA, U, US); Cuyuni-Mazaruni Region, Kuroba, E on Mazaruni R., 3 km N of confluence of Kamarang R., Clarke 808 (LEA). Suriname: Tumuc Humac Mts., Talouakem, upper Litani R., 10 km from first base camp, Acevedo 5827 (LEA); Para, Zanderij I, recreation area, 0.5 km E of Zanderij Hwy, 0.5 km N of town, Evans 1806 (MO). French Guiana: St. Elie, ORSTOM Biological Station at Forest Concession, Acevedo 4883 (F, LEA, MO, US); R. Comté, near Route N-2, Billiet & Jadin 1245 (BM, U).

Vernacular names: Suriname: fowroedoti = fouroedottie, pikien vouroedotie = pikien fowroedotie, koelatawéte = kurata weti.

Note: *P. crassifolium* is an extremely distinctive species because of its multiple, fertile intercalary cataphylls. While Eichler (1868) refers to and illustrates both the 2-seriate and 3-seriate condition, all specimens seen from western S America and C America appear to be strictly 2-seriate; however, an occasional third series of flowers has been noted in some Brazilian plants.

7. **Phoradendron dipterum** Eichler in Mart., Fl. Bras. 5(2): 109. 1868. Type: Brazil, Ceará, Gardner 1672 (holotype W, destroyed, isotypes BM, GH, K, P). – Fig. 32

Phoradendron demerarae Trel., Phoradendron 73. 1916. Type: Guyana, Demerara, Essequibo above Bartica, Jenman 2546 (holotype K, isotypes F, K, NY, U).

Glabrous, percurrent, often vigorous plants; internodes strongly compressed, to 8 cm long, some tending to be quadrangular or even 4-winged; lowest internodes of lateral shoots often keeled or more or less terete-quadrangular; basal cataphylls 1 pair, 0.3-0.5 cm above base, median in position, somewhat flaring when dry, rarely with a second pair somewhat higher. Leaf blade narrowly obovate to slightly spatulate, to 9(-13) x 3 cm, apex rounded, base tapering into 0.3-0.4(0.7) cm wide, flat, indistinct petiole, or somewhat amplexicaul; venation evident, of several basal, parallel veins nearly running into apex. Dioecious. Female inflorescence to 8.5 cm long, peduncle simple, 0.3-0.4 cm long, followed by about 6 fertile internodes; flowers 13-21 per fertile bract, 3-seriate. Fruit spherical, 0.2 x 0.2 cm, perianth segments closed.

Distribution: Tropical America, from C Mexico, Greater Antilles, to eastern Bolivia, Paraguay and Argentina; 250 collections studied (GU: 1; SU: 2).

Specimens examined: Guyana: Demerara, Essequibo above Bartica, Jenman 2546 (F, K, NY, U). Suriname: Forests of Zanderij, Samuels 451 (GH, K); near Paramaribo, Splitgerber 240 (L).

Note: *P. dipterum* is a member of a (perhaps always) hyperparasitic group of species which attack other mistletoes, often (but not exclusively) other species of *Phoradendron*. Its great variability of leaf size and stem morphology may be related to the different hosts parasitized, or perhaps to their vigor. The expression of the winged

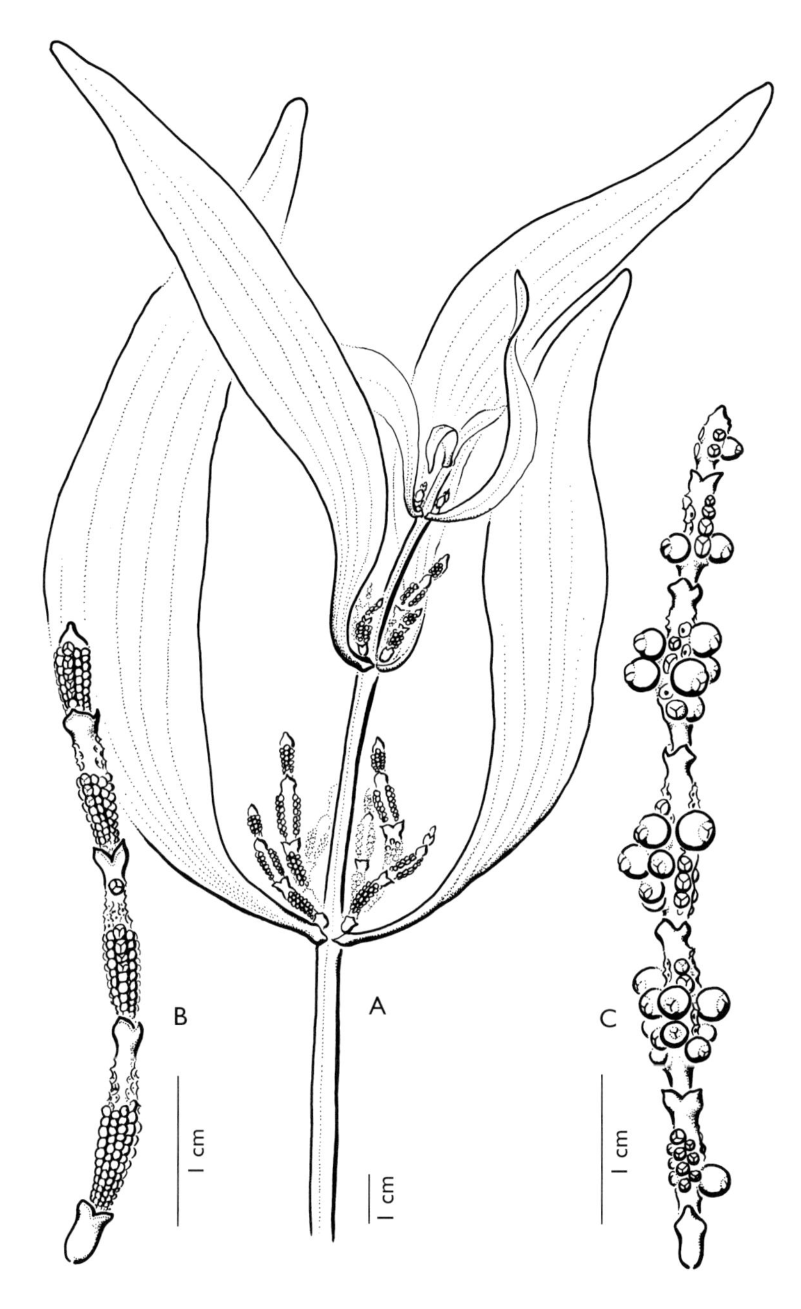

Fig. 32. *Phoradendron dipterum* Eichler: A, habit; B, male inflorescence; C, infructescence (reproduced from Kuijt, Syst. Bot. Monogr. 66, 2003).

character of the stem varies not only from plant to plant, but even within a plant; the lowest lateral internodes are usually wingless, and elsewhere 2-winged and 4-winged internodes are sometimes found on the same individual. Another interesting feature of the group is the fact that secondary stems, lacking basal cataphylls, may sprout from a basal cushion, not found in other species.

8. **Phoradendron granvillei** Kuijt, Syst. Bot. Monogr. 66: 219. 2003. Type: French Guiana, de Granville *et al.* 10802 (holotype CAY, isotypes B, LEA, NY, P, US). – Fig. 33

Internodes sharply keeled, to 5 cm long; basal cataphylls 1 pair, very low (0.2-0.3 cm), apparently soon deciduous; all scale leaves with conspicuous, light-colored margin. Petiole ca. 0.3 cm long, strongly winged; blade very thin, lanceolate, to 6.5 x 3 cm, apex acute or nearly so, base long-tapering into very indistinct petiole; venation pinnate but lowest lateral veins strong. Dioecious. Female inflorescence ca. 1.5 cm long at anthesis, elongating to 3-3.5 cm, extremely slender; peduncle 0.2-0.3 cm long; 2 sterile internodes, lower one very short, fertile internodes 5-8; flower 1 per fertile internode, placed in middle at anthesis, rarely with 1 or 2 additional flowers in lateral series, lower portion of internode elongating, placing fruit at distal end. Fruit white, ovoid, 0.25-0.3 x 0.15-0.2 cm, perianth segments closed.

Distribution: Known from the type only (FG: 1).

Note: Similar, and closely related to, *P. laxiflorum* Ule from the eastern Amazonian area, but immediately different in having no intercalary cataphylls. Additionally, *P. laxiflorum* has a very different leaf shape and venation, and its petals are erect (Kuijt 1986: 31). The male inflorescence of *P. laxiflorum* has 5-9 flowers per fertile bract in either 2- or 3-seriate pattern, even on the same plant, and we may expect a similar male inflorescence in *P. granvillei*.

9. **Phoradendron hexastichum** (DC.) Griseb., Fl. Brit. W. I. 313. 1860. – *Viscum hexastichum* DC., Prodr. 4: 282. 1830. Type: Cuba, De la Ossa s.n. (holotype G-DC). – Fig. 34

Phoradendron hexastichum (DC.) Griseb. var. *longispica* Eichler in Mart., Fl. Bras. 5(2): 129. 1868. Type: Brazil, Amazonas, Spruce 2112 (holotype M, isotypes BM, GH, K, NY, P).

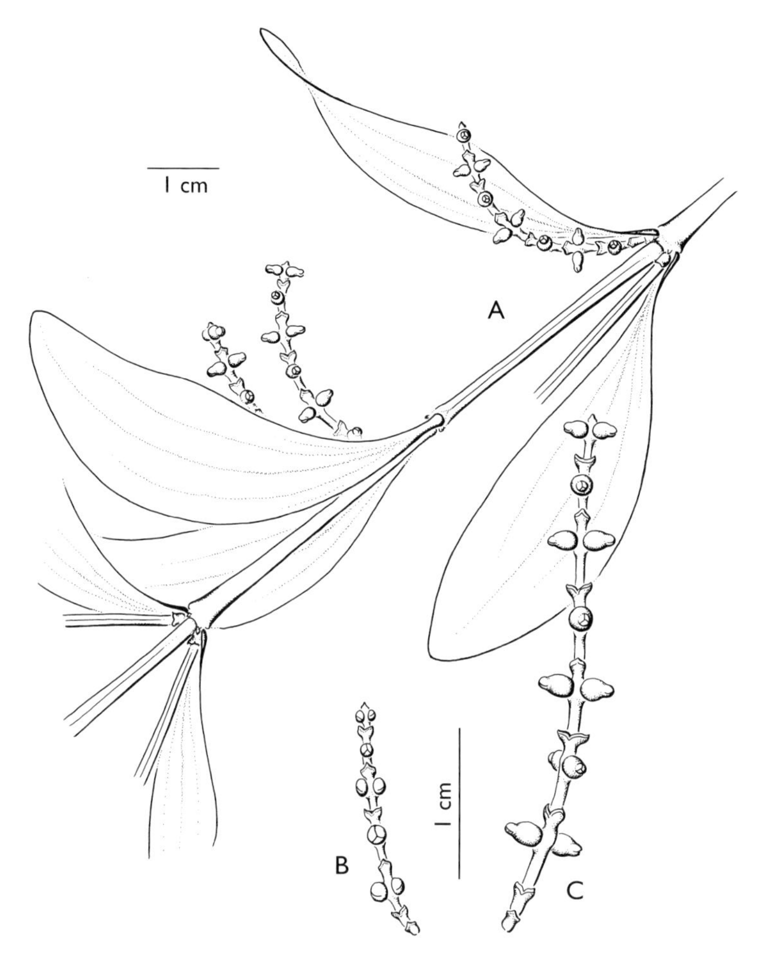

Fig. 33. *Phoradendron granvillei* Kuijt: A, habit; B, female inflorescence; C, infructescence (reproduced from Kuijt, Syst. Bot. Monogr. 66, 2003).

Rather large plants; internodes keeled to slightly quadrangular, becoming terete, to 10 cm long; basal cataphylls mostly 1 pair 0.3-0.4 cm above base, sometimes with a second, higher pair. Petiole 1-1.5 cm long; blade broadly ovate to obovate, 7-15(18) x 3-5(7) cm, apex rounded, base tapering to abruptly contracted; venation pinnate, midrib often strikingly light in color. Monoecious. Inflorescence 3-6.5 cm long, some inflorescences apparently entirely female, most mixed, with male flowers below female ones; peduncle 0.3-0.5 cm, simple or double, fertile internodes 3-5; flowers 6-10 per fertile bract, 3-seriate. Fruit white, globose to obovoid, 0.3 x 0.3 cm, perianth segments closed.

Fig. 34. *Phoradendron hexastichum* (DC.) Griseb.: A, habit of narrow-leaved S American variant; B, leaf of broad-leaved Caribbean variant (reproduced from Kuijt, Syst. Bot. Monogr. 66, 2003).

D i s t r i b u t i o n: Guerrero, Puebla, Veracruz, Belize, Costa Rica, Panama, Caribbean, S America to Brazil and eastern Bolivia; 250+ collections studied (GU: 2).

S p e c i m e n s e x a m i n e d: Guyana: Potaro-Siparuni Region, Kato, 2 km W of village, northern Pakaraima Mts., Mutchnik *et al.* 1383 (LEA); Kato Village, Mutchnik 1527 (LEA).

10. **Phoradendron inaequidentatum** Rusby, Bull. Torrey Bot. Club 27: 137. 1900. – *Dendrophtora inaequidentata* (Rusby) Trel., Phoradendron 218. 1916. Type: Bolivia, Rusby 1544 (holotype NY, isotypes F, GH, US). – Fig. 35

Phoradendron jenmanii Trel., Phoradendron 85. 1916. Type: Guyana, Demerara, Jenman 2541 (holotype K).
Phoradendron prancei Rizzini, Revista Fac. Agron. (Maracay) 8: 89. 1975. Type: Brazil, Amazonas, Allen *et al.* 2713 (holotype RB, isotypes MO, K, R, U, US).

Much branched plants with terete branches; nodes expanding in age; stems mostly dichotomous through terminal inflorescences, where percurrent with 1-4 pairs of intercalary cataphylls; lateral stems with 2-5 pairs of basal cataphylls spaced further apart distally but less than halfway to next foliage leaves. Petiole stout, to 1 cm long; blade elliptical to somewhat spatulate, to 9 x 6 cm; venation sometimes obscure, palmate. Monoecious. Inflorescence to 3 cm long when in fruit, bisexual; male flowers few and apparently limited to lowest 1(2) fertile internodes; sterile cataphylls 1-3 pairs, terminal inflorescence apparently without; fertile internodes 4 or 5; flowers 1-3 per fertile bract. Fruit somewhat reddish or orange, ovoid, nearly 0.4 x 0.6 cm, perianth segments erect.

D i s t r i b u t i o n: Panama to Bolivia; 150+ collections studied (GU: 8; FG: 1).

S e l e c t e d s p e c i m e n s: Guyana: Upper Takutu-Upper Essequibo Region, Rupununi R., between Kwattamang Landing and Rewa Village, Clarke *et al.* 6797 (LEA, US); near Kaieteur Falls, Cowan & Soderstrom 2076 (US); mile 87, Bartica-Potaro road, Fanshawe 2738 (K, US); Kaieteur Plateau, Maguire & Fanshawe 23134 (GH, K, US); Cuyuni-Mazaruni Region, Pakaraima Mts., NE plateau on Mt. Ayanganna, Henkel *et al.* 59 (LEA); Kamoa R., Toucan Mt., Jansen-Jacobs *et al.* 1632 (LEA, MO, U, US); Rupununi Distr., Kanuku Mts., Two-Head Mt., Jansen-Jacobs *et al.* 3590 (LEA, U). French Guiana: Mt. Bellevue de l'Inini, de Granville *et al.* 8026 (LEA, U).

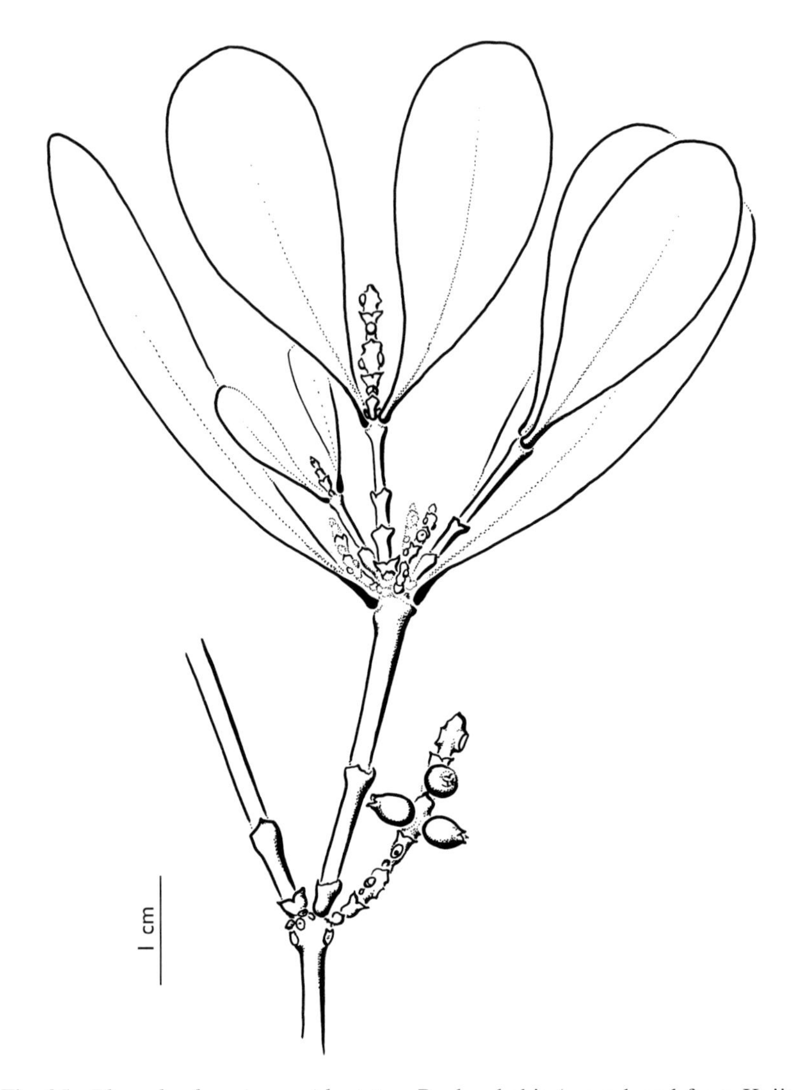

Fig. 35. *Phoradendron inaequidentatum* Rusby: habit (reproduced from Kuijt, Syst. Bot. Monogr. 66, 2003).

Note: Apparently, entirely male specimens occur, which show to 12 flowers per fertile bract in 3-seriate arrangement.

11. **Phoradendron krameri** Kuijt, Syst. Bot. Monogr. 66: 255. 2003. Type: Suriname, 20 km S of Paramaribo, Palisadenweg 2, Kramer & Hekking 2770 (holotype U). – Fig. 36

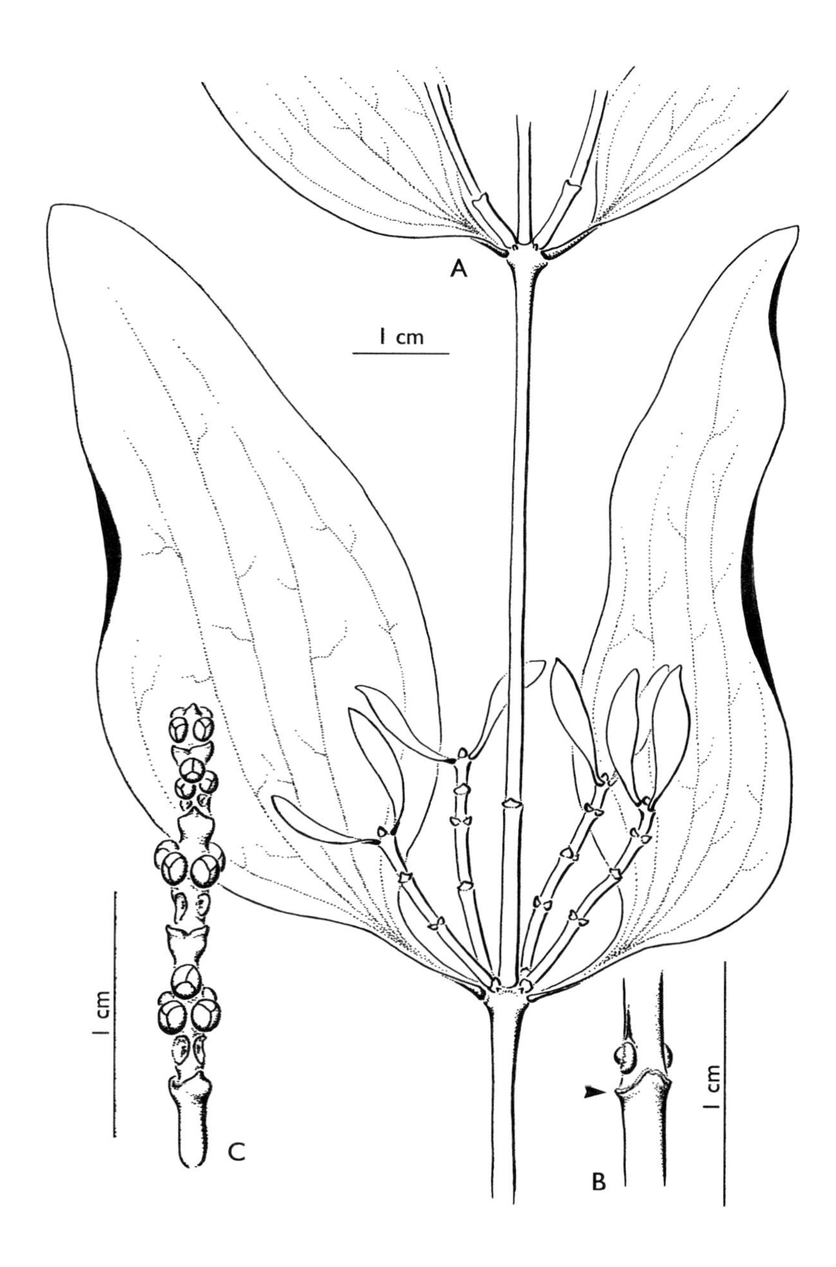

Fig. 36. *Phoradendron krameri* Kuijt: A, habit; B, node showing scars of intercalary cataphylls (arrow) and subtended inflorescence buds; C, inflorescence (reproduced from Kuijt, Syst. Bot. Monogr. 66, 2003).

Plants ca. 60 cm tall, bronze colored at least when fresh; intercalary cataphylls to 2.5 cm above foliar node, these sometimes fertile, or 2 pairs to 1 and 6 cm high; stem segments between foliar nodes to 10 cm, terete; basal cataphylls 1-4 pairs as high as 2 cm; all cataphylls caducous or eroding early, leaving inconspicuous scars. Leaf blade thin, ovate, to 10 x 3 cm, apex tapered but ultimately obtuse, base contracted to short, cuneate petiole; venation palmate, with 3 or 5 major veins. Monoecious. Male flowers few, in lower positions of some fertile internodes only. Inflorescence slender,1.5-2 cm long; peduncle short, simple or unequally double, fertile internodes mostly 4; flowers 3-5 per fertile bract, 2-seriate. Fruit (immature) with closed perianth segments.

Distribution: Guyana and Suriname; 3 collections studied (GU: 1; SU: 2).

Specimens examined: Guyana: Kalacoon, Jenman 2532 (K). Suriname: the type; between Coppename R. and Coronie, along Coronie road on shell ridge, Lanjouw & Lindeman 1466 (K, U).

12. **Phoradendron mairaryense** Ule in Pilg., Notizbl. Bot. Gart. Berlin-Dahlem 6: 291. 1915. Type: Brazil, Amazonas, Ule 8383 (holotype B, destroyed, isotypes K, L). – Fig. 37

Phoradendron sulfuratum Rizzini, Revista Fac. Agron. (Maracay) 8: 89. 1975. Type: Venezuela, Bolívar, Steyermark *et al.* 104137 (holotype RB, isotypes MO, NY).
Phoradendron linguiforme Rizzini in Maguire *et al.*, Mem. New York Bot. Gard. 29: 33. 1978. Type: Brazil, Amapá, Pires *et al.* 50445 (holotype RB, isotypes K, NY).

Large plants of yellowish color; internodes terete, to 6 cm long; basal cataphylls rather small, 2-4 pairs, upper pair to 3 cm above base; both percurrent and dichotomous; many branches eventually terminating in an inflorescence. Petiole 0.4 cm long; blade coriaceous, narrowly oblong, to 14 x 3 cm, apex rounded to nearly acute, base tapering into indistinct petiole; 3-5 palmate veins. Inflorescence to 4.5 cm long; peduncle simple or double, fertile internodes 4 or 5; flowers 3-5(7) per bract, 2-seriate (occasionally 3-seriate).

Distribution: Venezuela and Guyana into Brazil (Bahia); 24 collections studied (GU: 1).

Specimen examined: Guyana: Atkinson Field, Graham 526 (K).

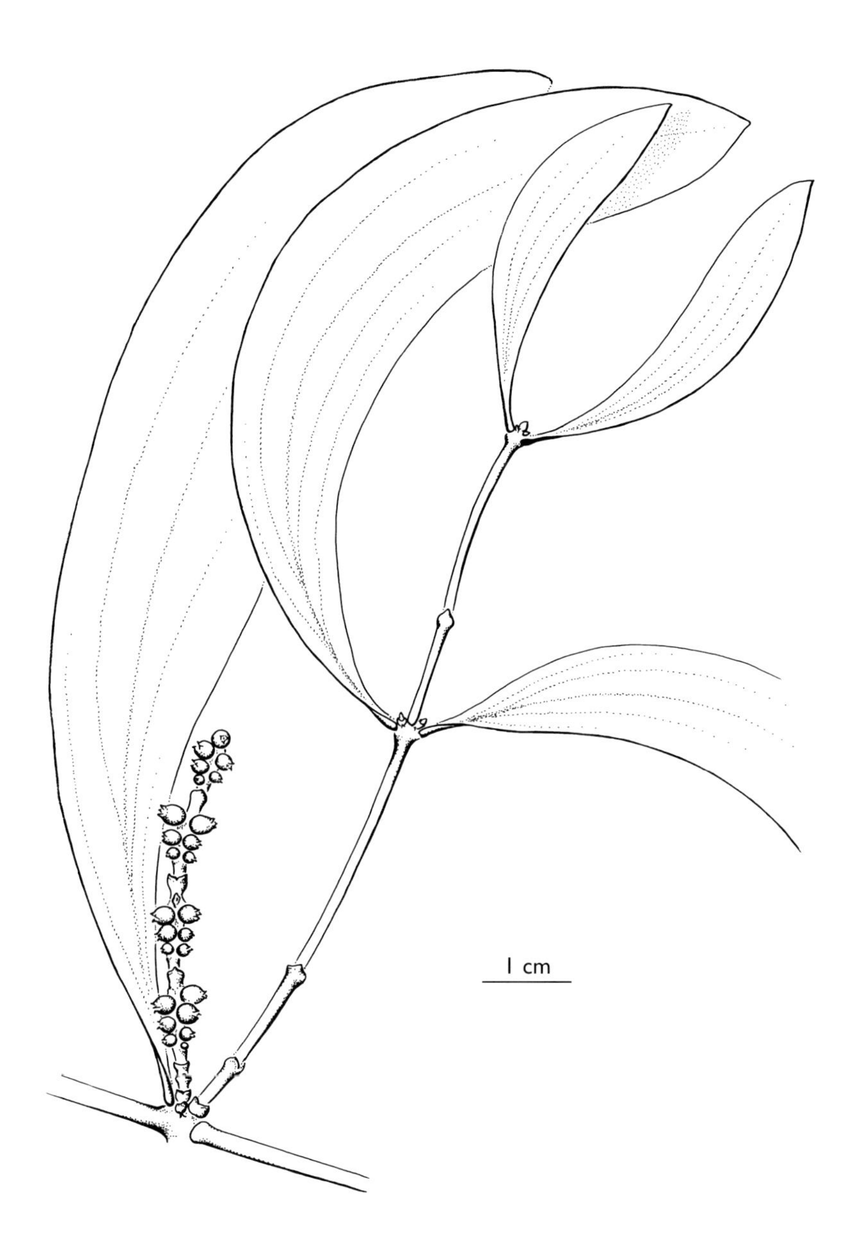

Fig. 37. *Phoradendron mairaryense* Ule: habit, the intercalary cataphylls shown should be regarded as exceptional (reproduced from Kuijt, Syst. Bot. Monogr. 66, 2003).

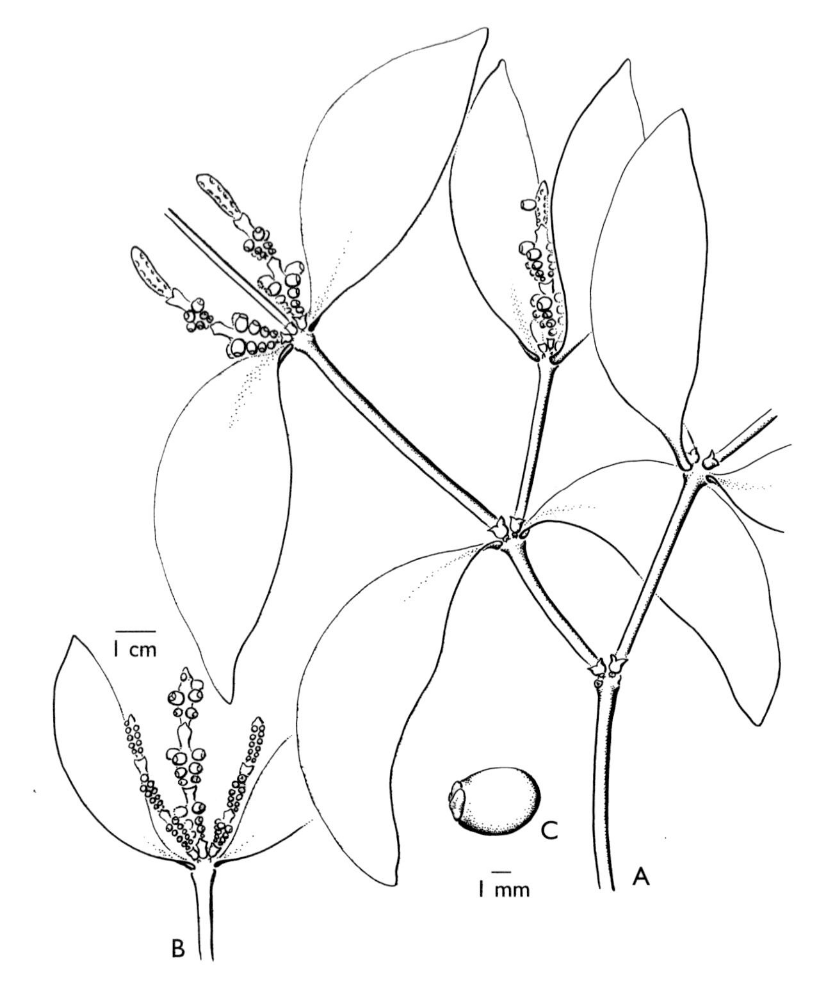

Fig. 38. *Phoradendron morsicatum* Rizzini: A, habit; B, young innovation with terminal inflorescence; C, fruit (reproduced from Kuijt, Syst. Bot. Monogr. 66, 2003).

13. **Phoradendron morsicatum** Rizzini in Maguire *et al.*, Mem. New York Bot. Gard. 29: 33. 1978. Type: Guyana, Upper Mazaruni R. basin, Mt. Ayanganna, on shoulder of E flank, above Thompson Camp, Tillett *et al.* 45088 (holotype RB, isotypes MO, NY, R).
– Fig. 38

Orange-yellow to yellowish brown plants, dichotomous through terminal inflorescences, fleshy; internodes terete, to 9 cm long; basal cataphylls 1 or 2 pairs, as low as 0.4 cm when only 1 pair, second pair to 1 cm above axil. Leaf blade leathery, lanceolate-ovate, to 9 x 5 cm, base acute, tapering into very short, poorly defined petiole; venation palmate but nearly obscure. Monoecious? Inflorescence to 4.5 cm long; peduncle usually simple (both in terminal and lateral inflorescences), rarely with a small pair of cataphylls, 5-7 cm long, fertile internodes 2 or 3; flowers 5-9 per fertile bract, 2-seriate. Fruit bright orange or brownish orange at maturity, 0.5 x 0.3 cm, perianth segments closed.

Distribution: Venezuela (Gran Sabana) and Guyana; 12 collections studied (GU: 10).

Selected specimens: Guyana: the type; Potaro-Siparuni Region, summit of Mt. Kopinang, Hahn *et al.* 4382 (LEA); Upland savanna 2-5 km N of Imbaimadai settlement, white sand savanna, Henkel 46 (LEA); Cuyuni-Mazaruni Region, Pakaraima Mts., Mt. Aymatoi, Maas *et al.* 5763 (K, LEA, MO, NY, U); Mazaruni-Potaro Region, Upper Mazaruni R. basin, N of Karowrieng R., peak of Pakaraima Mts., Pipoly *et al.* 7766 (BR, G, LEA, NY, US); Pakaraima Mts., headwaters of Mazaruni R., banks W of Imbaimadai, Pipoly *et al.* 7916 (LEA, NY, US); Ayanganna Plateau, savanna N of mountain center, Pipoly *et al.* 10899 (LEA, NY, US).

Note: Very similar to *P. berteroanum* (DC.) Griseb., differing in the low basal cataphylls, the very leathery leaves, and in inflorescence characters. *Phoradendron berteroanum*, previously known as *P. dichotomum* Krug & Urb., is not known from the Guianas.

14. **Phoradendron mucronatum** (DC.) Krug & Urb. in Urb., Bot. Jahrb. Syst. 24: 34. 1897. – *Viscum mucronatum* DC., Prodr. 4: 282. 1830. Type: Dominican Republic, Bertero s.n. (holotype G-DC). – Fig. 39

Densely leafy plants; young internodes quadrangular, mostly 2-5 cm long; basal cataphylls 1 pair ca. 0.2 cm above axil, inconspicuous. Petiole 0.2-0.3 cm long; blade mostly elliptical to obovate, to 3 x 1.5 cm, with rounded to mucronate or emarginate apex, base acute, tapering into indistinct petiole; venation of 3 conspicuous palmate veins. Monoecious. Inflorescence solitary or clustered at older nodes, ca. 2 cm at anthesis, elongation to 4.5 cm or less; peduncle simple, 0.1-0.2 cm long, fertile internodes (2)3-6(7); flowers 3 per fertile bract, in 3 series, apical one male, 2 lateral ones female. Fruit ovoid, 0.4 x 0.3 cm, with smooth or warty upper surface, perianth segments erect.

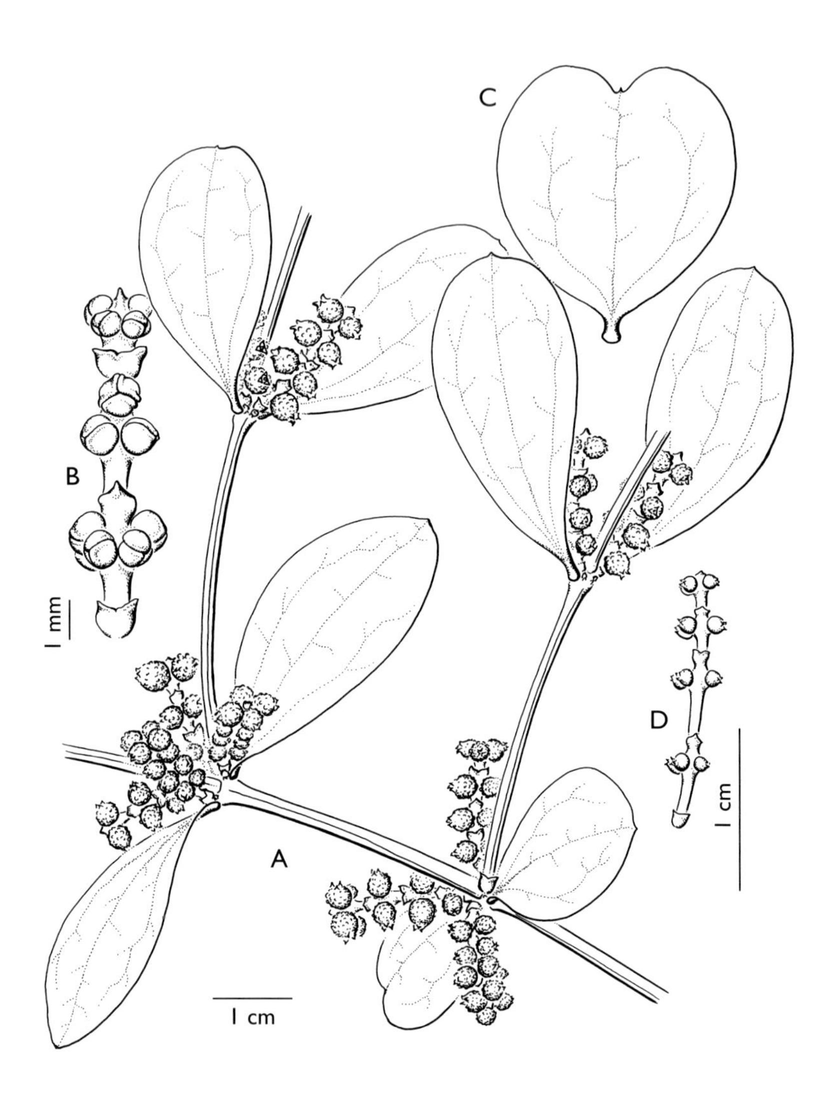

Fig. 39. *Phoradendron mucronatum* (DC.) Krug & Urb.: A, habit of fruiting plant; B, inflorescence, only the apical flowers of internodes male; C, notched leaf type; D, immature inflorescence of variant with elongated inflorescence axes (reproduced from Kuijt, Syst. Bot. Monogr. 66, 2003).

Distribution: Mexico (Queretaro, San Luis Potosi, Yucatan), Panama, Caribbean, northern S America, to amazonian Bolivia and Paraguay; 300+ collections studied (GU: 5).

Specimens examined: Guyana: Upper Takutu-Upper Essequibo Region, Rupununi Savanna, Tup-Tup Ialli, 4 km W of Wichabai, Gillespie 1961 (K, LEA, MO, NY, U, US); Rupununi Distr., Simony Cr., Görts-van Rijn *et al.* 347 (LEA); E bank of Takutu R., S of Lethem, Irwin 815 (US); Kanuku Mts., Rupununi R., Bush Mouth near Witaru Falls, Jansen-Jacobs *et al.* 130 (LEA, U, US); SE of Karasabai to Yourora Cr., 3 km from water tower, McDowell 2194 (HUA, LEA, US).

Note: An extremely variable species with, consequently, a very complex synonymy (Kuijt 2003: 305-306). Many, but by no means all, of the continental plants have much longer inflorescences than the Caribbean material, and the warty fruit surface seems less pronounced, although herbarium material provides weak information on this point.

15. **Phoradendron northropiae** Urb. in Northr., Mem. Torrey Bot. Club 12: 33, f. 4. 1902. Type: Bahamas, Andros Island, Northrop & Northrop 551 (holotype F, isotypes GH, K, NY). – Fig. 40

Dichotomous, erect plants; internodes terete but somewhat keeled below young nodes, to 7 cm long; basal cataphylls 1 pair, somewhat tubular, 0.2-0.3 cm above base, inconspicuous. Petiole ca. 0.3 cm long ; blade obovate, to 6 x 3.5 cm, apex rounded to notched, base contracted to indistinct; venation pinnate but obscure. Monoecious. Inflorescence slender, to nearly 2 cm; peduncle 0.3-0.5 cm, with 1 short internode followed by 1 twice as long; fertile internodes mostly 2 or 3; flowers 5-9 per fertile bract, 2-seriate. Fruit white.

Distribution: A mistletoe of extremely spotty occurrence: Bahamas, Venezuela (Amazonas), Guyana, Brazil, amazonian Peru and Bolivia; 15 collections studied (GU: 2).

Specimens examined: Guyana: Essequibo, Ikwaka Lake, Allison 149 (K); Mazaruni-Potaro, Essequibo R., Moroballi Cr., near Bartica, Sandwith 551 (K, RB, U).

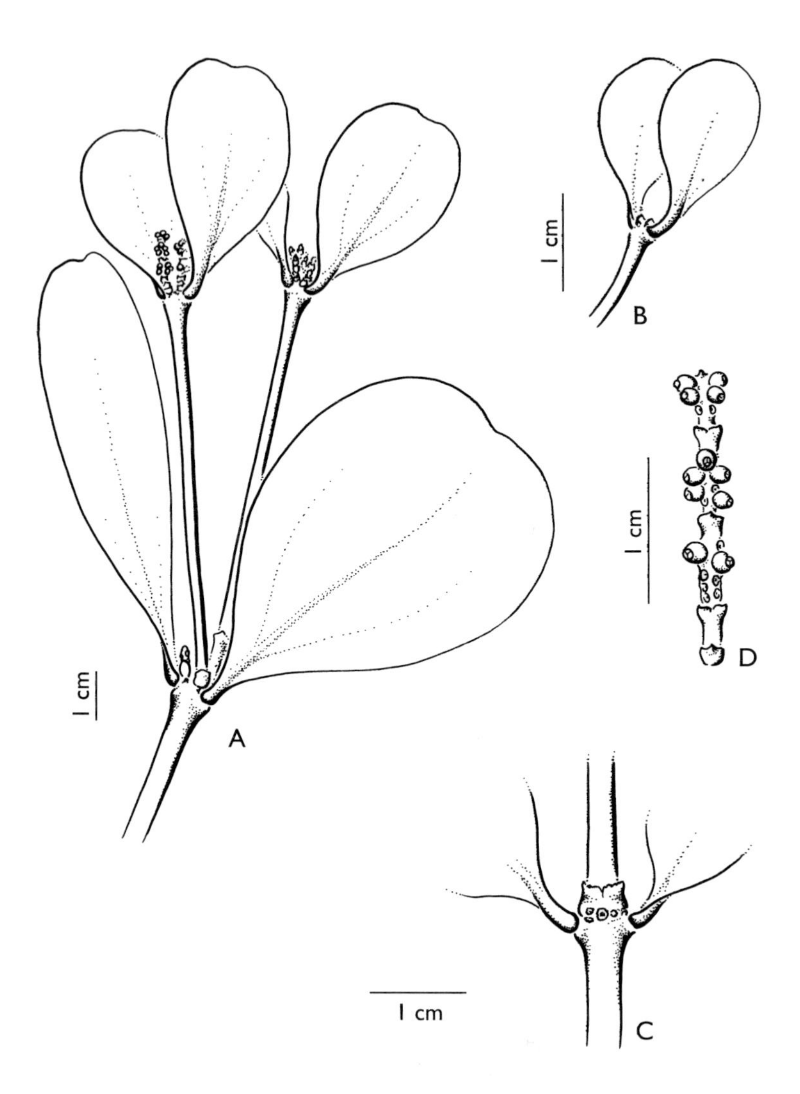

Fig. 40. *Phoradendron northropiae* Urb.: A, habit; B, aborting shoot tip; C, single branch of dichotomy resembling percurrent growth; D, infructescence (reproduced from Kuijt, Syst. Bot. Monogr. 66, 2003).

Note: The following specimen was collected on the Guyana-Brazil boundary, and cannot be placed securely in either country: Acarai Mts., height of land between drainage of Rio Mapuera (Trombetas tributary) and Shodikar Cr. (Essequibo tributary), A.C. Smith 2967 (A, F, K, MO, U, US).

16. **Phoradendron obtusissimum** (Miq.) Eichler in Mart., Fl. Bras. 5(2): 134m. 1868. – *Viscum obtusissimum* Miq., Linnaea 18: 602. 1845 ('1844'). Type: Suriname, Para, Focke 1019 (holotype U, isotype K). – Fig. 41

Phoradendron acinacifolium Mart. ex Eichler in Mart., Fl. Bras. 5(2): 117. 1868. Type: Brazil, Rio de Janeiro, Gaudichaud 574 (lectotype P, isolectotype G, designated by Kuijt 1994: 189).
Phoradendron acinacifolium Mart. ex Eichler var. *surinamense* Rizzini in Maguire *et al.*, Mem. New York Bot. Gard. 29: 32. 1978. Type: Suriname, Maguire *et al.* 54147 (holotype NY, isotypes K, MO, RB, US).

Plants percurrent; internodes compressed/keeled above, terete below, 3-8 cm long; basal cataphylls 1 pair, 0.1-0.3 cm above base, very inconspicuous. Petiole ca. 0.5 cm long, rather slender; blade obovate or oblong, to 8 x 3 cm, apex rounded, base cuneate, tapering into indistinct petiole; venation often with 2 prominent, long lateral veins. Monoecious. Inflorescence with variable sex distribution: some fertile internodes or even spikes entirely male, others mixed, usually with 3 apical flowers of a flower area female; flowers 4-9 per fertile bract, 2-seriate; 1 extremely low (0.1 cm) pair of sterile bracts, elongating from 1 cm at anthesis to twice that length in fruit; fertile internodes 2 or 3; fruits 3 above each fertile bract, placed very high on internode. Fruit translucent white, 0.6-0.7 x 0.35 cm, perianth segments large, flaring.

Distribution: Costa Rica through tropical S America to Paraguay, Argentina, and eastern Bolivia; 120+ collections studied (GU: 12; SU: 2).

Selected specimens: Guyana: Kaieteur Plateau, Kaieteur Falls to Mure-mure Savanna, Cowan & Soderstrom 2129 (F, K, NY, US); Rupununi Distr., Massara, Graham 255 (K); Potaro-Siparuni Region, Kaieteur Falls, along river above falls, Hahn *et al.* 4579 (LEA, NY, U, US); Pakaraima Mts., Kurupung R., Imatta Cr. near Kurupung Landing, Hoffman *et al.* 2401 (HUA, LEA, U, US); Cuyuni-Mazaruni Region, Pakaraima Mts., Imbaimadai Cr., 1 km W of Imbaimadai, Hoffman *et al.* 1631 (LEA); Rupununi Savanna, Mora Savanna near Toroebaroc Cr., Jansen-Jacobs *et al.* 1105 (K, LEA, NY, U).

17. **Phoradendron pellucidulum** Eichler in Mart., Fl. Bras. 5(2): 106. 1868. Type: Brazil, Amazonas, Spruce 3480 (holotype W, destroyed, isotypes K, P). – Fig. 42

Phoradendron semivenosum Rizzini, Rodriguésia 18-19(30-31): 189. 1956. Type: Venezuela, Bolívar, Steyermark 59388 (holotype VEN).

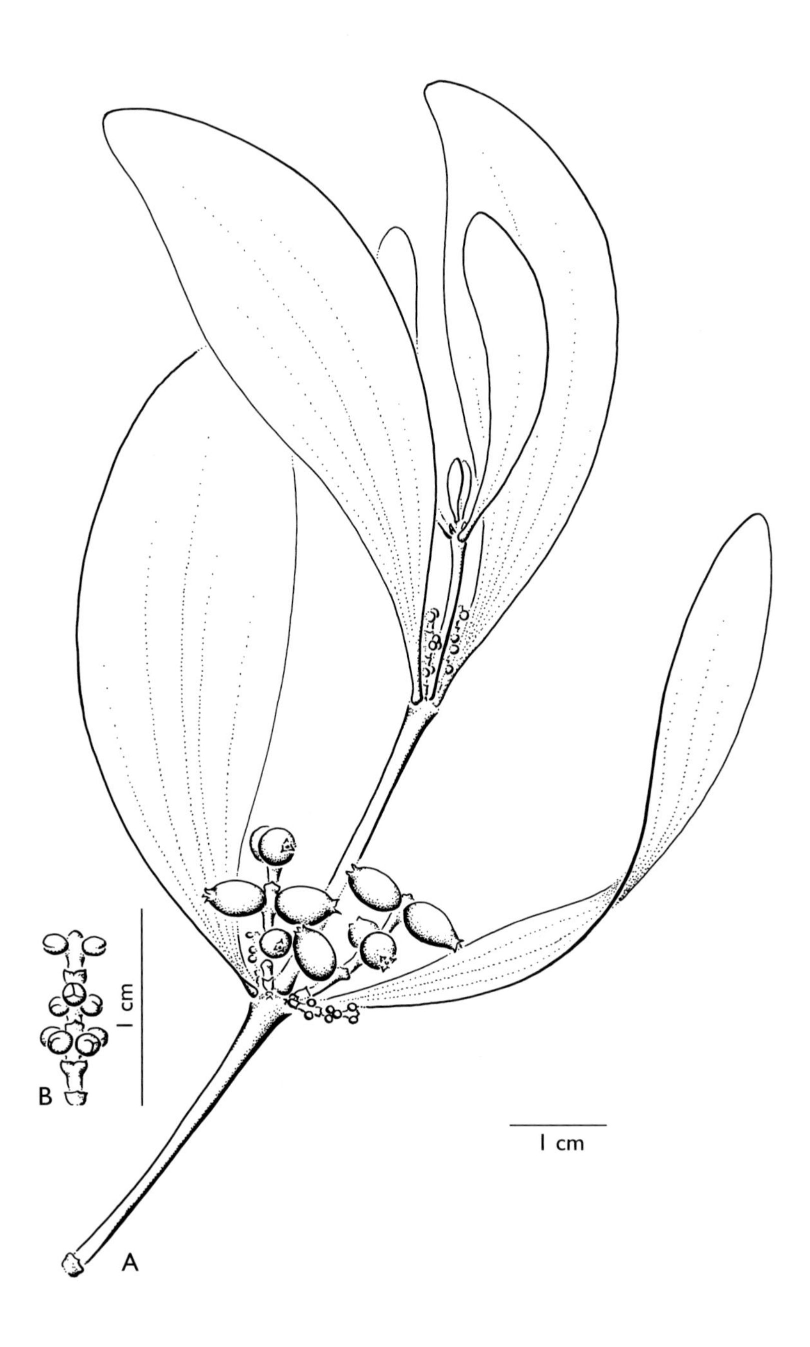

Fig. 41. *Phoradendron obtusissimum* (Miq.) Eichler: A, habit of fruiting plant; B, inflorescence (reproduced from Kuijt, Syst. Bot. Monogr. 66, 2003).

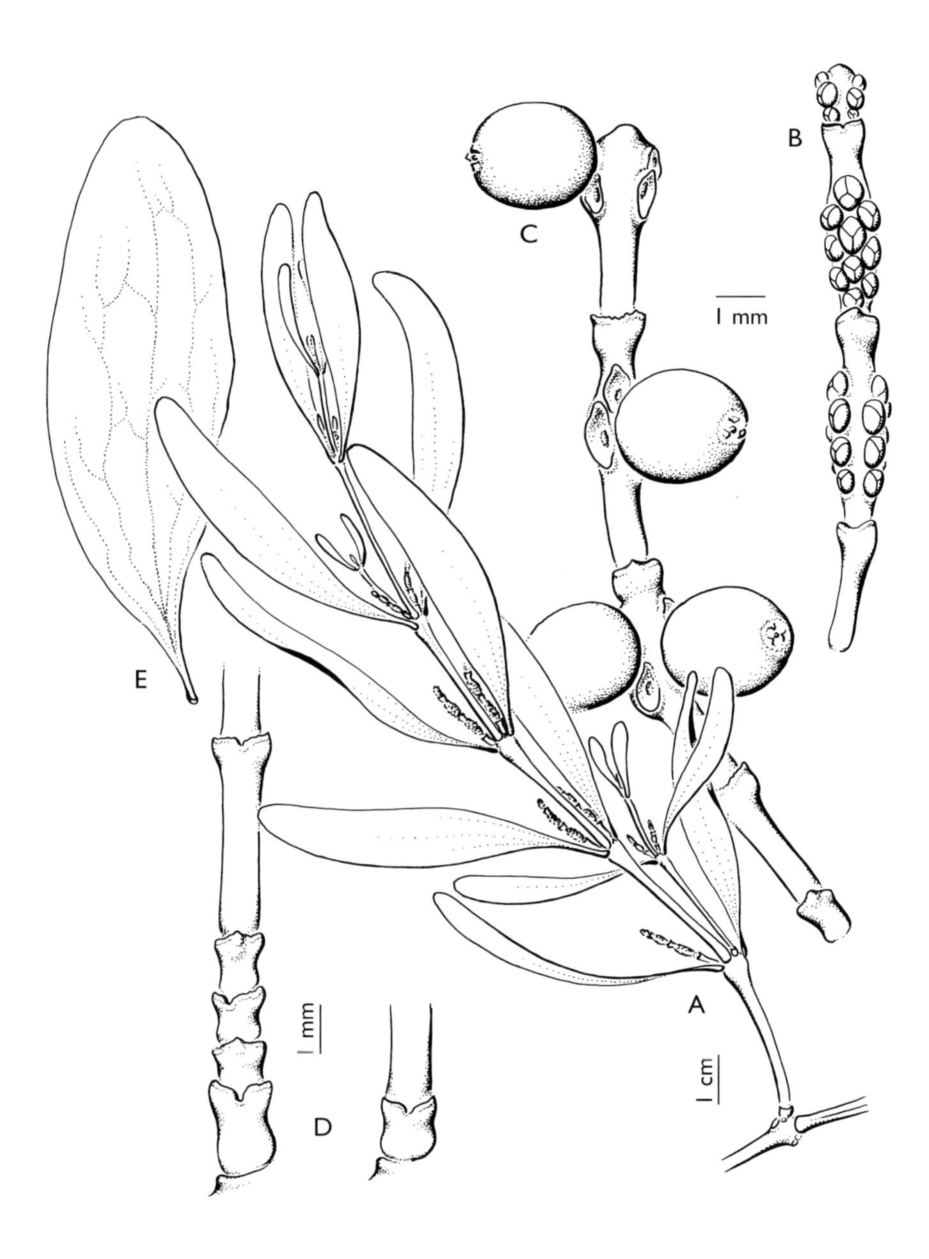

Fig. 42. *Phoradendron pellucidulum* Eichler: A, habit; B, male inflorescence; C, infructescence; D, basal cataphylls; E, large leaf (reproduced from Kuijt, Syst. Bot. Monogr. 66, 2003).

Phoradendron williamsii Rizzini, Rodriguésia 18-19(30-31): 188. 1956. Type: Venezuela, Alto Orinoco, Williams 15429 (holotype VEN, isotypes F, RB, US).
Phoradendron insigne Steyerm. in Steyerm. *et al.*, Bol. Soc. Venez. Ci. Nat. 26: 415. 1966. Type: Venezuela, Bolívar, Steyermark & Aristiguieta 59 (holotype VEN, isotypes F, K, NY, U).
Phoradendron pseudomucronatum Rizzini, Rodriguésia 28(41): 18. 1976. Type: Venezuela, Bolívar, Steyermark *et al.* 104167 (holotype RB).
Phoradendron semivenosum var. *agostinii* Rizzini, Rodriguésia 30(46): 86. 1978. Type: Venezuela, Bolívar, Agostini 398 (holotype US, isotypes K, U, US).
Phoradendron semivenosum var. *longipes* Rizzini, Rodriguésia 30(46): 86. 1978. Type: Venezuela, Bolívar, Agostini 400 (holotype US).

Delicate, percurrent, much branched plants; leaf-bearing internodes 2-5 cm long, orange brown, terete and ca. 0.1 cm thick below when leaf-bearing, expanding somewhat and slightly keeled above; basal cataphylls 1-5. Leaf blade thin, narrowly elliptical, mostly to 6 x 1.5 cm, apex rounded, with long, slenderly tapering base; venation inconspicuous, pinnate, often with 2 strong basal veins. Dioecious; occasional bisexual plant and possibly some unisexual inflorescences. Inflorescence peduncle mostly simple but sometimes with to 2 pairs of sterile cataphylls, 0.1-0.3 cm long; fertile internodes 2 or 3. Male inflorescence to 1.5 cm long; flowers to 18 per fertile bract, 3-seriate. Female inflorescence to 2 cm long in fruit; flowers mostly 3 per fertile bract. Fruit globose, 0.3 cm in diameter, perianth segments closed.

Distribution: Venezuela (Amazonas, Bolívar) and Guyana to southern Brazil and C Paraguay; 125+ collections studied (GU: 7).

Specimens examined: Guyana: Paruima-Utshi R., trail to Venezuela junction, Clarke 862 (LEA); Mazaruni-Potaro, Imbaimadai, Partang R. crossing, 1 km from mouth, Gillespie *et al.* 2691 (LEA, U, US); Pakaraima, Kuraku burru, Graham 535 (K); Pakaraima Mts., Heika R., 4 km from Chinoweing Village, Hoffman *et al.* 3363 (US); Rupununi Distr., Kuyuwini Landing, Kuyuwini R., Jansen-Jacobs *et al.* 2976 (K, LEA, U); Upper Mazaruni R. basin, Karowrieng R., Pakaraima Mts., unnamed peak NW of Maipuri Falls, Pipoly *et al.* 7688 (LEA, US, NY); Imbaimadai Cr., W of Imbaimadai, Pipoly *et al.* 7960 (LEA, P, U, US).

Note: Superficially somewhat reminiscent of a slender form of *P. quadrangulare*, but mostly dioecious, and with very different cataphyll characteristics. Also similar to slender plants of *P. obtusissimum* in general aspect but, in contrast to that species, mostly dioecious and with simple peduncles. Young leaves characteristically tend to be reddish-translucent at least when dry.

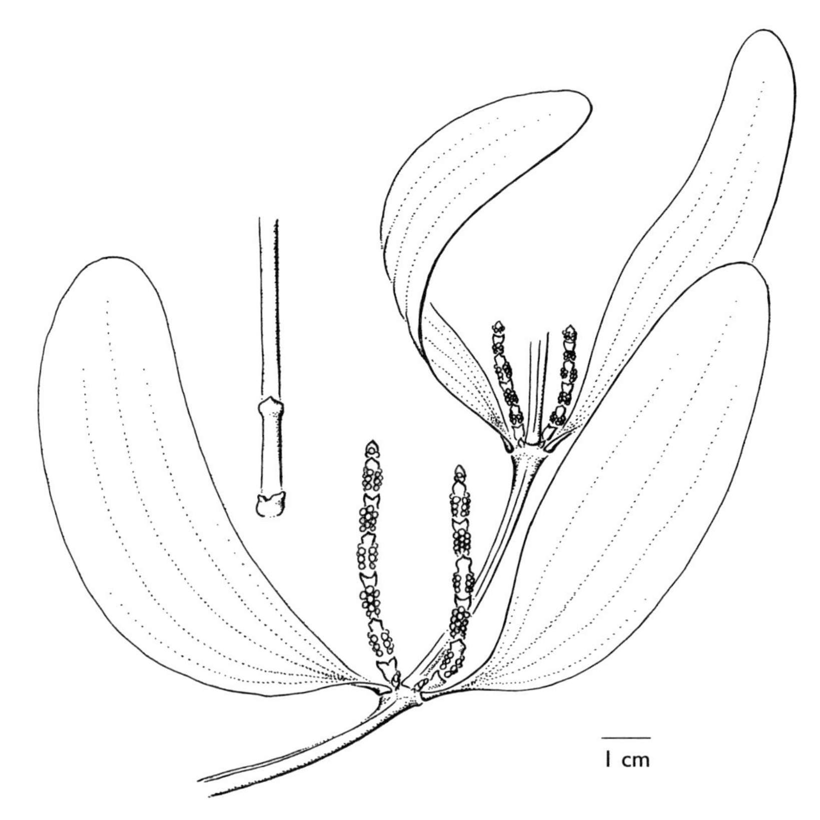

Fig. 43. *Phoradendron perrottetii* (DC.) Eichler: habit and basal cataphylls (reproduced from Kuijt, Syst. Bot. Monogr. 66, 2003).

18. **Phoradendron perrottetii** (DC.) Eichler in Mart., Fl. Bras. 5(2): 112. 1868. – *Viscum perrottetii* DC., Prodr. 4: 280. 1830. Type: French Guiana, Perrottet 228 (holotype G-DC). – Fig. 43

Viscum dimidiatum Miq., Linnaea 18: 58. 1844. – *Phoradendron dimidiatum* (Miq.) Eichler in Mart., Fl. Bras. 5(2): 134m. 1868. Type: Suriname, Focke 716 (holotype U, isotype K).
Phoradendron harmsianum Ule in Pilg., Notizbl. Bot. Gart. Berlin-Dahlem 6: 290. 1915. Type: Brazil, Amazonas, Ule 7889 (holotype B, destroyed, see F Negative # 18188).

Usually rather large plants; internodes terete or slightly keeled above. to 9 cm long; intercalary cataphylls absent (rarely 1 pair at about 1.5 cm); basal cataphylls 1 or 2(3) pairs, lowest at about 0.6 cm, highest to 2 cm above base. Petiole 0.5 cm long; blade elliptical or dimidiate to falcate, to

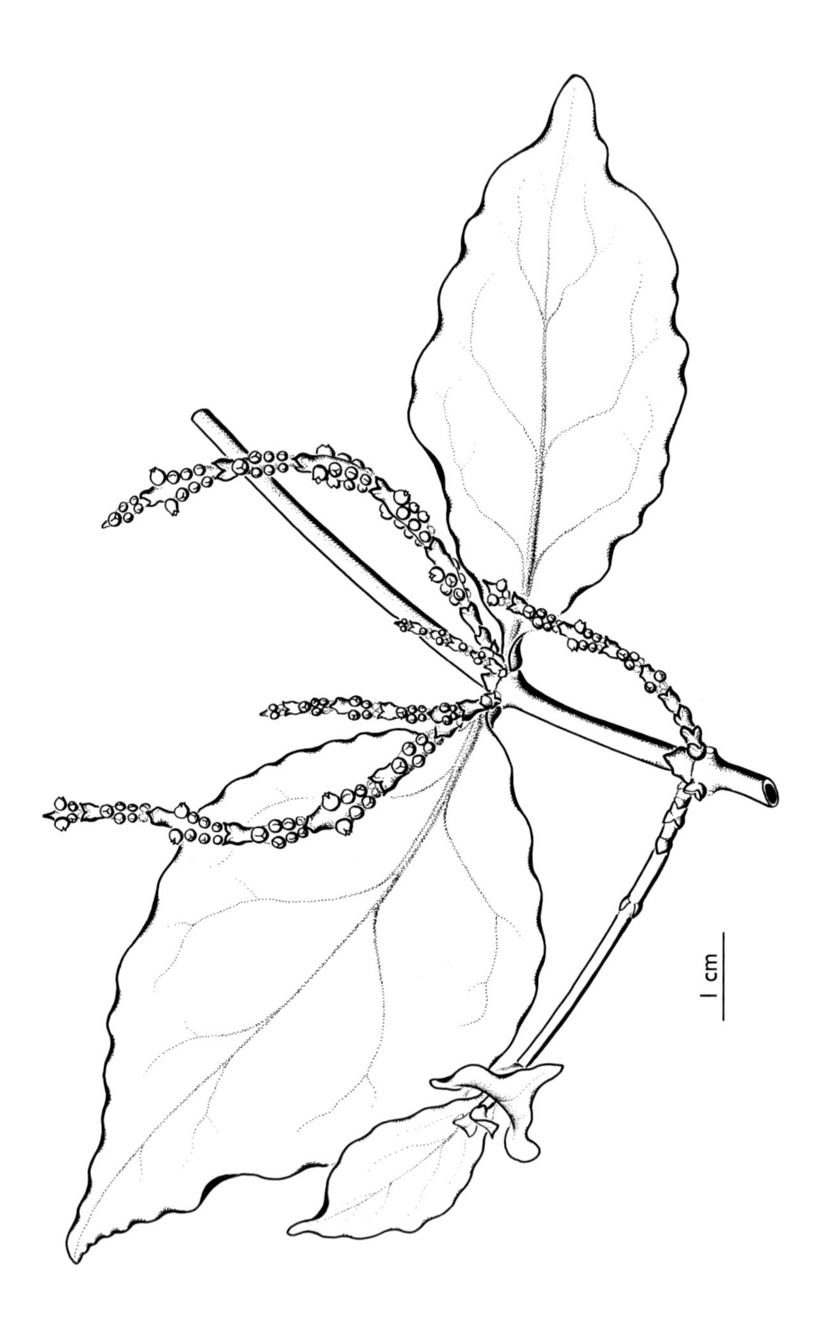

Fig. 44. *Phoradendron piperoides* (Kunth) Trel.: habit (reproduced from Kuijt, Syst. Bot. Monogr. 66, 2003).

15 x 4 cm, apex rounded, base cuneately tapering into indistinct petiole; venation pinnate, but basal lateral veins often very strong. Monoecious. Inflorescence often long and sinuous, 8 cm long or longer with 10 or more fertile internodes; sterile cataphylls 0-2 pairs, very low; fertile internodes with to 7-22 flowers per bract in mostly 3-seriate pattern, female flowers sometimes formed only on 1 side of internode or in 1 series. Fruit translucent white, spherical, 0.4 x 0.3 cm, perianth segments closed.

Distribution: Venezuela to eastern Bolivia; 230+ collections studied (GU: 31; SU: 7; FG: 6).

Selected specimens: Guyana: Essequibo, Pomeroon Distr., Mora Landing, Moruka R., de la Cruz 1848 (F, GH, MO, US); Mazaruni-Potaro, Bartica, Jenman 4821 (BM, K); Pakaraima Mts., Imbaimadai Cr., W of Imbaimadai, Pipoly *et al.* 7998a (U, US); Rupununi, Surama Village, dirt road to main road, Acevedo *et al.* 3403 (F, K, LEA, NY, US). Suriname: La Campagne, in Bellanakreek, Focke 716 (K, U); near Paramaribo, Focke 168 (L); Jodensavanne-Mapane Kreek area, Suriname R., near Camp 8, Schulz 7266 (K, MO, U). French Guiana: Route de l'Anse, from Sinnemary to Battures de Malmanoury, Billiet & Jadin 1732 (LEA).

Vernacular names: Suriname: fowroedoti, pikien foroeka = pikin fowrukaka, kantakamma.

19. **Phoradendron piperoides** (Kunth) Trel., Phoradendron 145. 1916. – *Loranthus piperoides* Kunth in Humb., Bonpl. & Kunth, Nov. Gen. Sp. ed. qu. 3: 443. 1820. – *Viscum piperoides* (Kunth) DC., Prodr. 4: 281. 1830. Type: Colombia, Humboldt & Bonpland s.n. (holotype P-Bonpl.). – Fig. 44

Viscum schottii Pohl ex DC., Prodr. 4: 281. 1830. – *Phoradendron schottii* (Pohl ex DC.) A. Gray, U.S. Explor. Exped., Bot. Phan. 742. 1854. Type: Brazil, Pohl s.n. (holotype G-DC).
Viscum fockeanum Miq., Linnaea 18: 60. 1844. Type: Suriname, Focke 281 (lectotype U, here designated).

Stems terete; successive pairs of foliage leaves separated by 1 pair of intercalary cataphylls 0.5-1 cm above node, leaves and cataphylls thus alternating; lateral branches with several pairs of basal cataphylls. Petiole 0.4-0.8 cm long; blade lanceolate, up to 14 x 7 cm, often with slightly attenuate apex and undulate margin, base acute. Monoecious; entirely unisexual plants do not seem to exist; number and positon of few male flowers variable, apical or scattered among females; often no more than 1 male flower above each bract, and sometimes none. Inflorescence up to 8 cm long; 5-8 fertile internodes above and 0-several pairs of sterile cataphylls below; flowers either 2-seriate or 3-seriate, sometimes on same spike, when distal internodes tending to be 2-seriate; each flower area with up to 15 flowers. Fruit yellow or yellowish-red, spherical to elliptical, ca. 0.3 cm in diam., perianth segments erect; seed very flat.

Distribution: Common and very widespread at lower elevations, from Mexico and the Caribbean south to Peru, Bolivia, Argentina and Paraguay; 800+ collections studied (FG: 26; GU: 40+; SU: 15).

Selected specimens: Guyana: Upper Takutu-Upper Essequibo Region, Makawatta massif, W part, summit of unnamed peak, Clarke 1851 (LEA, US); NW Distr., Amakura R., de la Cruz 3429 (F, GH, K, MICH, UC, US); Kanuku Mts., Rupununi R., Bush Mouth near Witaru Falls, Jansen-Jacobs *et al.* 77 (K, LEA, MO, U, US, WIS). Suriname: Tumuc Humac Mts., second peak S of Talouakem, Acevedo *et al.* 6104 (LEA); Saramacca Polder, Gonggrijp 2045 (U); Lower Suriname R., Plantation Peperpot, Lanjouw 580 (K, MO, S, U); Brownsberg Nature Park, 90 km S of Paramaribo, Mazaruni Plateau, Mori *et al.* 8402 (U). French Guiana: Mont Cabassou, Cayenne Island, Cremers *et al.* 14372 (LEA, U); Town of Maripasoula, Maroni basin, road to Maripasoula village, Fleury 433 (LEA); Monpé Soula, Haut-Marouini basin, de Granville *et al.* 10077 (LEA, NY, U, S); Haut Litany, Litany basin, de Granville *et al.* 12024 (LEA, NY, US).

Vernacular names: Suriname: fowroedoti, wérokarotika.

20. **Phoradendron poeppigii** (Tiegh.) Kuijt, Acta Bot. Neerl. 10: 199. 1961. – *Dendrophthora poeppigii* Tiegh., Bull. Soc. Bot. France 43: 182. 1896. Type: Brazil, Poeppig s.n. (holotype P). – Fig. 45

Squamate plants; green, compressed young internodes to 7 cm long and 3-4 mm wide, becoming terete; all long internodes, whether in percurrent or lateral positions, with 1 pair of short, brownish, tubular cataphylls 0.1-0.2 cm above node; innovations with 3-5 percurrent internodes eventually followed by a terminal inflorescence. Leaf scales 0.1 cm long, acute and apparently caducous. Monoecious. Inflorescence mostly lateral, 1 cm long, doubling in length when fruiting, with 1 sterile internode 0.1 cm long and 3 or 4 fertile internodes; flowers 3 per bract, apical one male, 2 lateral ones female. Fruit orange, ovoid, 0.6 x 0.5 cm, smooth, perianth segments erect.

Distribution: Venezuela, S Guyana, Brazil (Pará); 10 collections studied (GU: 1).

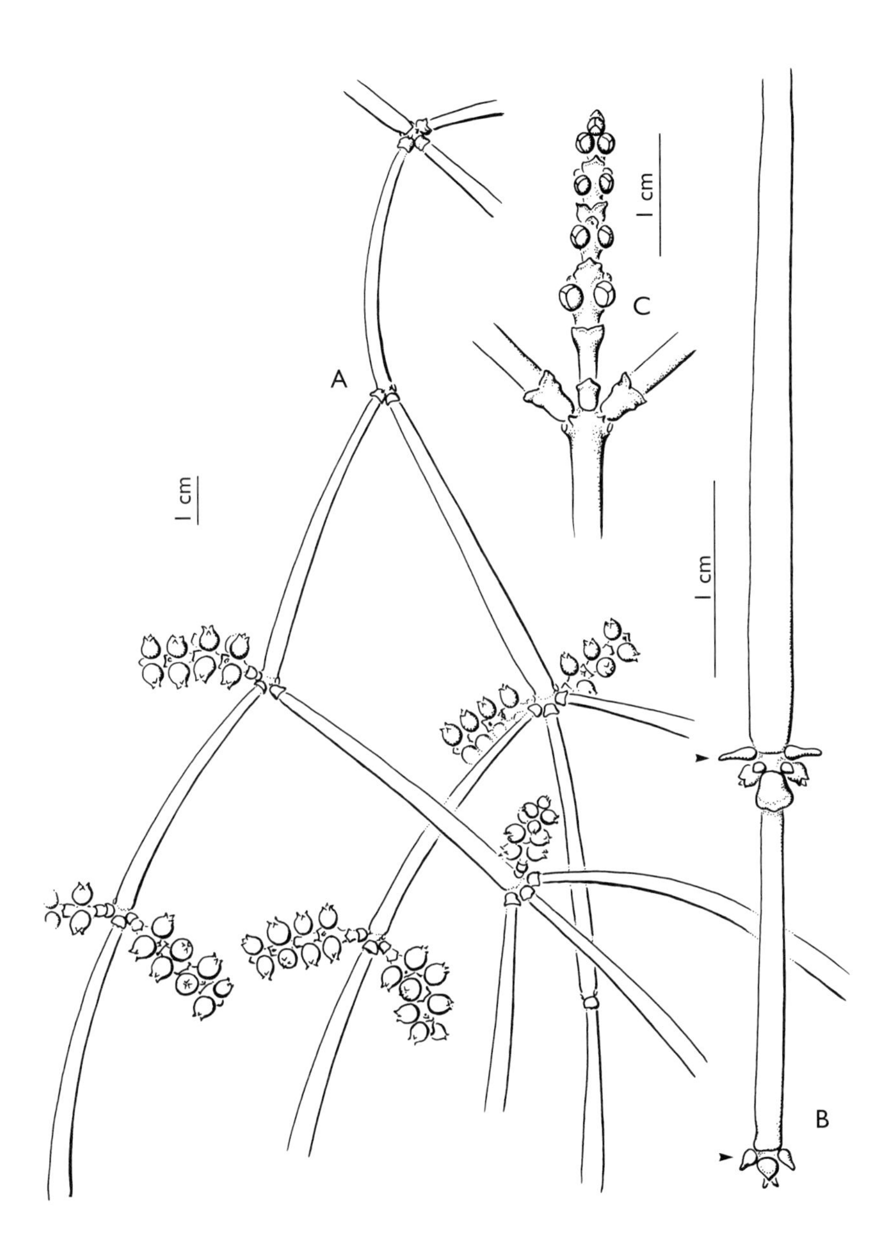

Fig. 45. *Phoradendron poeppigii* (Tiegh.) Kuijt: A, habit; B, youngest part of growing shoot, showing caducous leaves (arrows); C, terminal inflorescence (reproduced from Kuijt, Syst. Bot. Monogr. 66, 2003).

Specimen examined: Guyana: Rupununi Distr., Kuyuwini Landing, Kuyuwini R., Jansen-Jacobs *et al.* 2380 (K, LEA, MO, NY, U, US).

Note: The species is closely related to *P. platycaulon* Eichler, *P. tunaeforme* (DC.) Eichler, and *P. strongylocIados* Eichler.

21. **Phoradendron pteroneuron** Eichler in Mart., Fl. Bras. 5(2): 127. 1868. Type: Brazil, near Rio de Janeiro, Glaziou 1462 (lectotype BR, designated by Trelease 1916: 140). – Fig. 46

Young stems slightly keeled, soon becoming terete, percurrent; internodes to 7 cm long; intercalary cataphylls absent; laterals with 1 or 2(3) pairs of rather large basal cataphylls, to 2 cm or more above axil. Petiole ca. 1 cm long; blade (ob)ovate to elliptic, to 9(-12) x 6 cm, apex rounded, base abruptly tapered into petiole. Monoecious? (spikes of all 3 specimens identical). Inflorescence to 5 cm long, with 1-6 sterile internodes, fertile internodes 3-6, crowded; flowers (3)5(7) per fertile bract, 2-seriate. Fruit ovoid, 0.4 x 0.25 cm, smooth, perianth segments closed.

Distribution: Colombia, Venezuela, Guyana, Suriname, Brazil (Rio de Janeiro), Bolivia; 60+ collections studied (GU: 4; SU: 1).

Specimens examined: Guyana: Rupununi, trail Kopinang-Orinduik, ca. 1.5 mi walk from Orinduik, Boom *et al.* 9301 (K, LEA, NY, U, US); Mazaruni-Potaro, Parakaima Mts., Cipo Mts., sheetrock area N and NW of summit ridge, Henkel *et al.* 1142 (LEA); Ayanganna Plateau, Koatse R. valley, 2 km E of base camp, Pipoly *et al.* 10933 (LEA, NY, U, US); Headwaters of Kangu R., W branch, 4 km W of peak of Mt. Ayanganna, first talus slope of Plateau, Pipoly *et al.* 11010 (LEA, NY, U, US). Suriname: Sipaliwini savanna area on Brazilian frontier, Oldenburger *et al.* 814 (NY, U).

Note: Similar in general appearance, especially leaf shape and size, to *P. racemosum*, which ranges through the Guianas, the Caribbean, and northern Brazil. *P. racemosum*, however, is consistently dichotomous (through abortion of its apex), while *P. pteroneuron* is percurrent; and the inflorescence of the former has a peduncle of no more than 2 sterile internodes, while up to 6 are present in *P. pteroneuron*.

22. **Phoradendron pulleanum** K. Krause, Recueil Trav. Bot. Neérl. 22: 346. 1925. Type: Suriname, Brownsberg, BW (Boschwezen) 2489 (holotype U, isotypes NY, K, L, RB, US). – Fig. 47

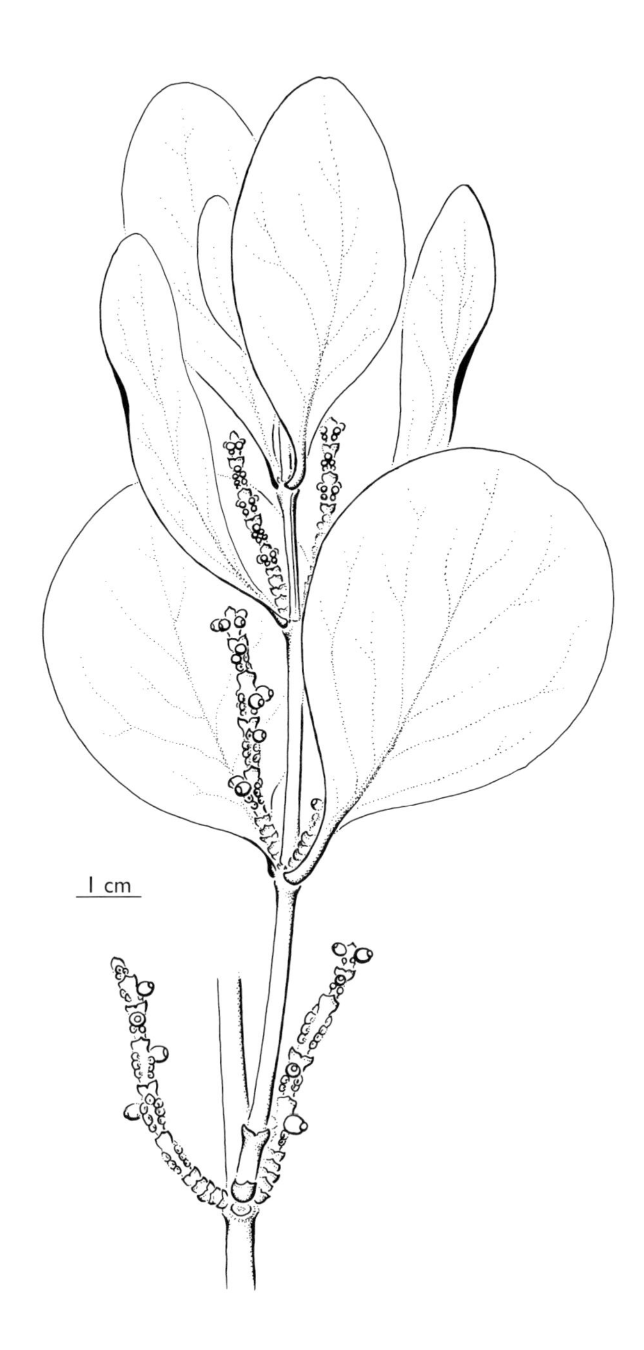

Fig. 46. *Phoradendron pteroneuron* Eichler: habit (reproduced from Kuijt, Syst. Bot. Monogr. 66, 2003).

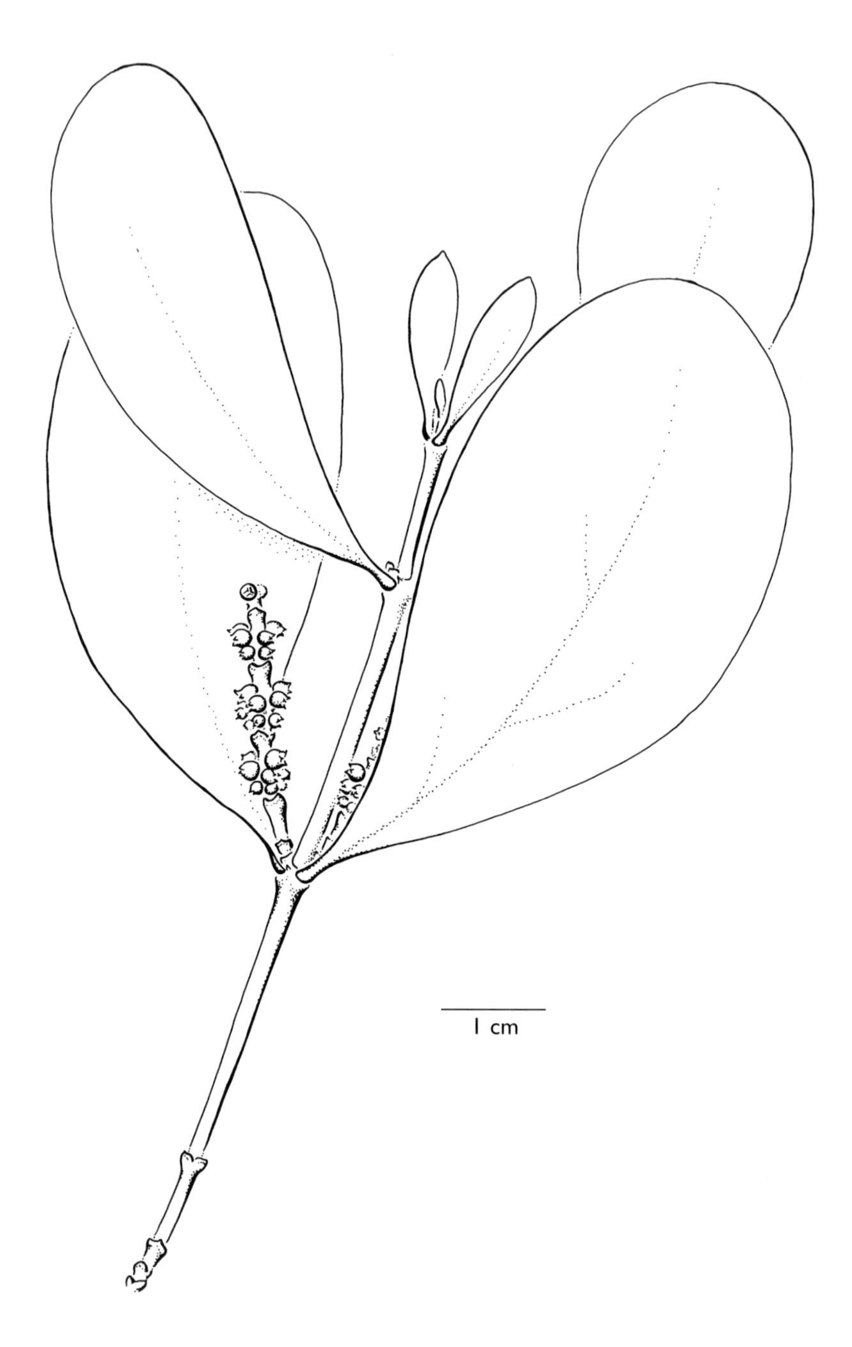

Fig. 47. *Phoradendron pulleanum* K. Krause: habit (reproduced from Kuijt, Syst. Bot. Monogr. 66, 2003).

Rather leggy, percurrent plant; internodes terete below, expanding just below massive node, to 7 cm long; basal cataphylls 1(2) pair, 1-1.5 cm above axil. Petiole ca. 0.5 cm long; blade coriaceous, narrowly obovate to spathulate, to 9 x 3 cm, apex rounded, base cuneately tapering into indistinct petiole; only midrib evident. Dioecious? Inflorescence to 3 cm long, with 1-4 sterile internodes and 3 or 4 fertile ones; flowers 5-9 per fertile bract, 2-seriate.

Distribution: Venezuela, the Guianas; 4 collections studied (GU: 1; SU: 2; FG: 1).

Specimens examined: Guyana: near Mabura Hill sawmill, along logging road under construction, Stoffers *et al.* 146 (LEA). Suriname: the type; Lely Mts., SW plateaux covered by ferrobauxite, beyond E end of airstrip, Lindeman & Stoffers *et al.* 812 (COL, F, MO, NY, U). French Guiana: Mt. Atachi Bacca, Region of Inini, NW summit, 6 km E of Gobaya Soula, vicinity of camp, de Granville *et al.* 10529 (LEA).

Vernacular name: Suriname: fowroedoti = fouroedottie.

23. **Phoradendron quadrangulare** (Kunth) Griseb., Fl. Brit. W. I. 711. 1864, as 'Phorodendron'. – *Loranthus quadrangularis* Kunth in Humb., Bonpl. & Kunth, Nov. Gen. Sp. ed. qu. 3: 444. 1820. – *Viscum quadrangulare* (Kunth) DC., Prodr. 4: 283. 1830. Type: Colombia, Humboldt & Bonpland s.n. (holotype P-Bonpl.). – Fig. 48

Phoradendron martianum Trel., Phoradendron 114. 1916. Type: Brazil, Gardner 1321 (holotype P).

Plants often rather small, glabrous, erect; internodes sharply quadrangular at least when young; basal cataphylls 1, nearly axillary, inconspicuous. Petiole indistinct; blade obovate to lanceolate-spatulate, 2-5 cm, apex rounded, base cuneate. Monoecious. Inflorescence to 2.5 cm or more, 1-3 in leaf axils, mostly lacking sterile bracts; fertile internodes 3-6; flowers mostly to 9 per fertile bract, 2-seriate. Fruit yellow or slightly orange (not red), spherical, 0.2-0.3 cm in diam., perianth segments closed.

Distribution: Probably the most widespread of all American mistletoes, ranging from Mexico, C America, the Greater Antilles, south to amazonian Peru and Bolivia and northern Paraguay and Argentina, surprisingly not yet recorded from Suriname or French Guiana; 1000+ collections studied (GU: 2).

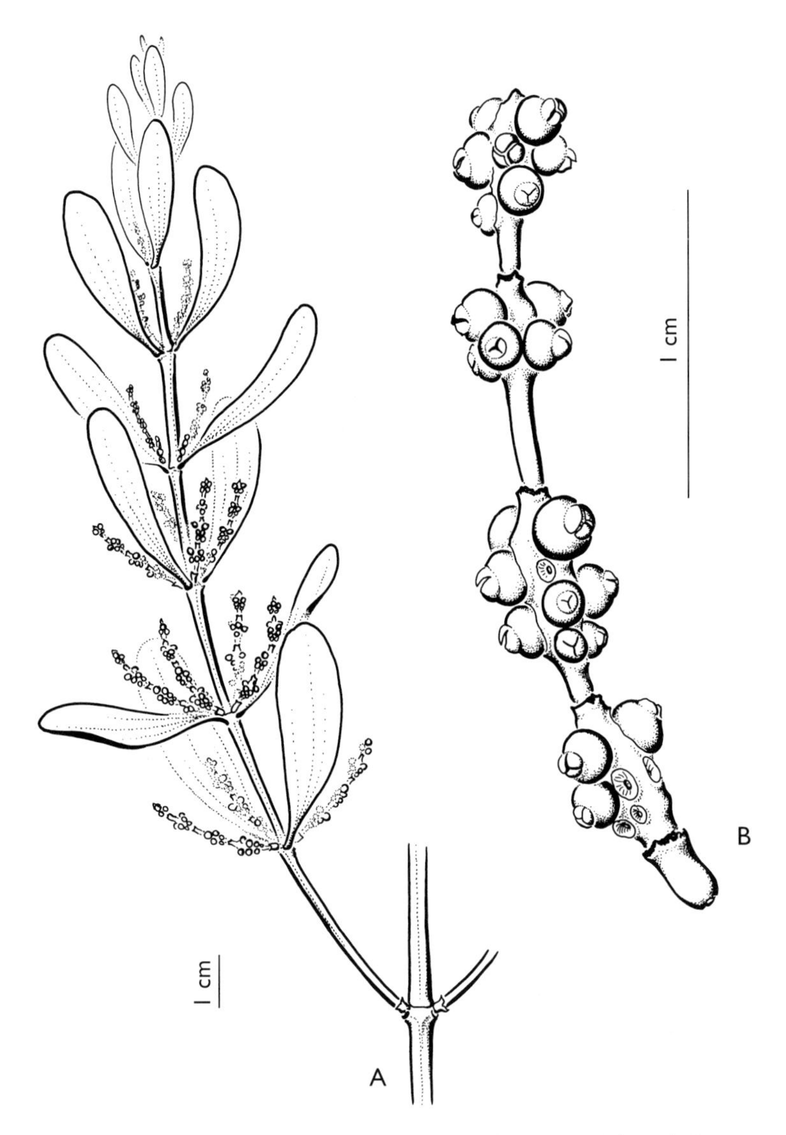

Fig. 48. *Phoradendron quadrangulare* (Kunth) Griseb.: A, habit; B, infructescence (reproduced from Kuijt, Syst. Bot. Monogr. 66, 2003).

Specimens examined: Guyana: along trail from Morris Mines (on Ireng R.) to Karasabai Village, along Karabaikura Cr., Knapp & Mallett 2882 (LEA); Rupununi Distr., Chaakoitou, near Mountain Point, just S of Kanuku Mts., Maas *et al*. 4074 (LEA, U).

Note: With respect to Rizzini's decision (1978: 82-84) to place *P. quadrangulare* in synonymy to *P. rubrum* (L.) Griseb., see the discussions in Kellogg & Howard (1986: 86-88) and Kuijt (2003: 369-376).

24. **Phoradendron racemosum** (Aubl.) Krug & Urb. in Urb., Bot. Jahrb. Syst. 24: 46. 1897. – *Viscum racemosum* Aubl., Hist. Pl. Guiane 2: 895. 1775. Type: [icon] Plum., Pl. Amer. ed. Burm. t. 258, f. 4. 1760. – Fig. 49

Viscum glandulosum Miq., Linnaea 18: 60. 1844. – *Phoradendron glandulosum* (Miq.) Eichler in Mart., Fl. Bras. 5(2): 134m. 1868. Type: Suriname, Para, Osembo, Focke 586 (holotype U, probable isotype Focke s.n. K).
Phoradendron cayennense Eichler in Mart., Fl. Bras. 5(2): 129. 1868. Type: French Guiana, Richard s.n. (holotype P).

Rather large plants often drying blackish, dichotomous by abortion; internodes quadrangular above, rather long (often 10-12 cm or more), with 1 or 2 pairs of basal cataphylls 0.5-1 cm above axil, where stem tends to be terete; nodes becoming swollen. Petiole 0.5-1 cm long; blade elliptical to ovate or broadly lanceolate, mostly to 12 x 8 cm, apex blunt to acute, base abruptly tapering into petiole. Monoecious. Inflorescences first paired, later clustered at nodes, mostly to 3 cm long, with 1 very low pair of cataphylls ca. 0.2 cm above base, rest of peduncle ca. 0.3 cm long, followed by 3 fertile internodes each with 3-9 flowers per bract in 2-seriate pattern, terminal 1-3 flowers of each flower area apparently male, others female. Fruit oblong, 0.25 x 0.15 cm, perianth segments small, closed.

Distribution: Bahamas, Costa Rica, Greater Antilles, Venezuela, the Guianas, Brazil; 200+ collections studied (GU: 25; SU: 6; FG: 5).

Selected specimens: Guyana: Rupununi, Kuywini R., 0-5 km W of landing at terminus of trail from Karaudarnau and Parabara Savanna, Clarke 5066 (LEA, US); Essequibo, Camaria Falls, Cuyuni R., Davis 1026 (K); Upper Mazaruni R., Kamakusa, de la Cruz 2833 (F, GH, MO, UC, US); Essequibo R., forest near Rockstone, Maas *et al.* 3944 (K, LEA, MO, U). Suriname: Bergendaal, Focke 989 (U); Tafelberg, S rim Arrowhead Basin, Maguire 24655 (F, GH, K, MO, U, US); Village Pikipada, Sauvain 217 (LEA, US). French Guiana: Mts. des Chevaux, near Route de l'Est, Gentry *et al.* 63182 (LEA); Mana R., Tamanoir Cr., Hallé 582 (US); Camp No. 3, Akouba Booka goo Soula, Bassin du Haut-Marouini, de Granville *et al.* 10015 (LEA, NY, U, US).

Vernacular names: Suriname: fowroedoti, piki fo kaka, wirikaro.

Fig. 49. *Phoradendron racemosum* (Aubl.) Krug & Urb.: A, habit; B, aborted shoot apex (reproduced from Kuijt, Syst. Bot. Monogr. 66, 2003).

25. **Phoradendron strongyloclados** Eichler in Mart., Fl. Bras. 5(2): 109. 1868. Type: Brazil, Pernambuco, Island of Itamavia, Gardner 1029 (holotype B, destroyed, isotypes BM, GH, K, NY, P). – Fig. 50

Phoradendron surinamense Pulle, Enum. Vasc. Pl. Surinam 155. 1906. Type: Suriname, Versteeg 239 (holotype U).
Phoradendron essequibense Trel., Phoradendron 149. 1916. Type: Guyana, Demerara, Jenman 2252 (holotype K, isotype NY).

Stems terete or slightly flattened below nodes; internodes to 8 cm long; 1 pair of prominent, yellowish basal cataphylls 0.2-0.3 cm above base of each lateral branch, and a similar pair at base of each percurrent leaf-bearing internode. Petiole 0.2 cm long; blade rather leathery, narrowly ovate, to 4 x 1 cm, apex attenuate, base contracted to petiole. Monoecious. Inflorescences paired or clustered at nodes but also terminal, about 1.5 cm long, lateral inflorescence mostly with 1 pair of sterile bracts followed by 0.1 cm peduncle and 3 fertile internodes with 3 flowers per fertile bract (1 on terminal internodes); male flowers in apical position above fertile bract and falling away very early, female ones in lateral series; terminal inflorescence of some plants with simple peduncle. Fruit yellow-orange to red, ovoid, 0.4 x 0.3 cm, perianth segments spreading.

Distribution: Venezuela, the Guianas, adjacent Brazil, to amazonian Bolivia; 50+ collections studied (GU: 15; SU: 4; FG: 4).

Selected specimens: Guyana: Ikwaka Lake, Essequibo, Allison 150 (K); E of Linden at junction of road to Mabura Hill and road to Bartica-Potaro road, Boom 7138 (LEA, U, US); Kamoa Mts., 2 km N of camp on Kamoa R., Clarke 2974 (LEA); Berbice R., Kibilibiri R., Forest Department F-59 (K); Potaro-Siparuni, Pakaraima Mts., upper Ireng R. watershed, E bank Kaatnang R., at base of Achiknang, Henkel *et al.* 5728 (LEA); Kanuku Mts., Nappi Mt., Jansen-Jacobs *et al.* 866 (LEA, K, MO, U). Suriname: Zanderij I, Gonggrijp s.n. (U); along railway near Republiek, Lanjouw & Lindeman 101 (K, U); Zanderij II, Stahel s.n. (U). French Guiana: vicinity of Cayenne, Broadway 612 (GH); Badreul, Broadway 722 (GH); small island opposite Antecume Pata, confluence of Litani (Haut Maroni) and Marouini Rs., Cremers 5061 (LEA).

26. **Phoradendron trinervium** (Lam.) Griseb., Fl. Brit. W. I. 314. 1860. – *Viscum trinervium* Lam., Encycl. 3: 57. 1789. Type: [icon] Plum., Pl. Amer. ed. Burm. t. 258, f. 2. 1760 (lectotype, designated by Kuijt 2003: 451). – Fig. 51

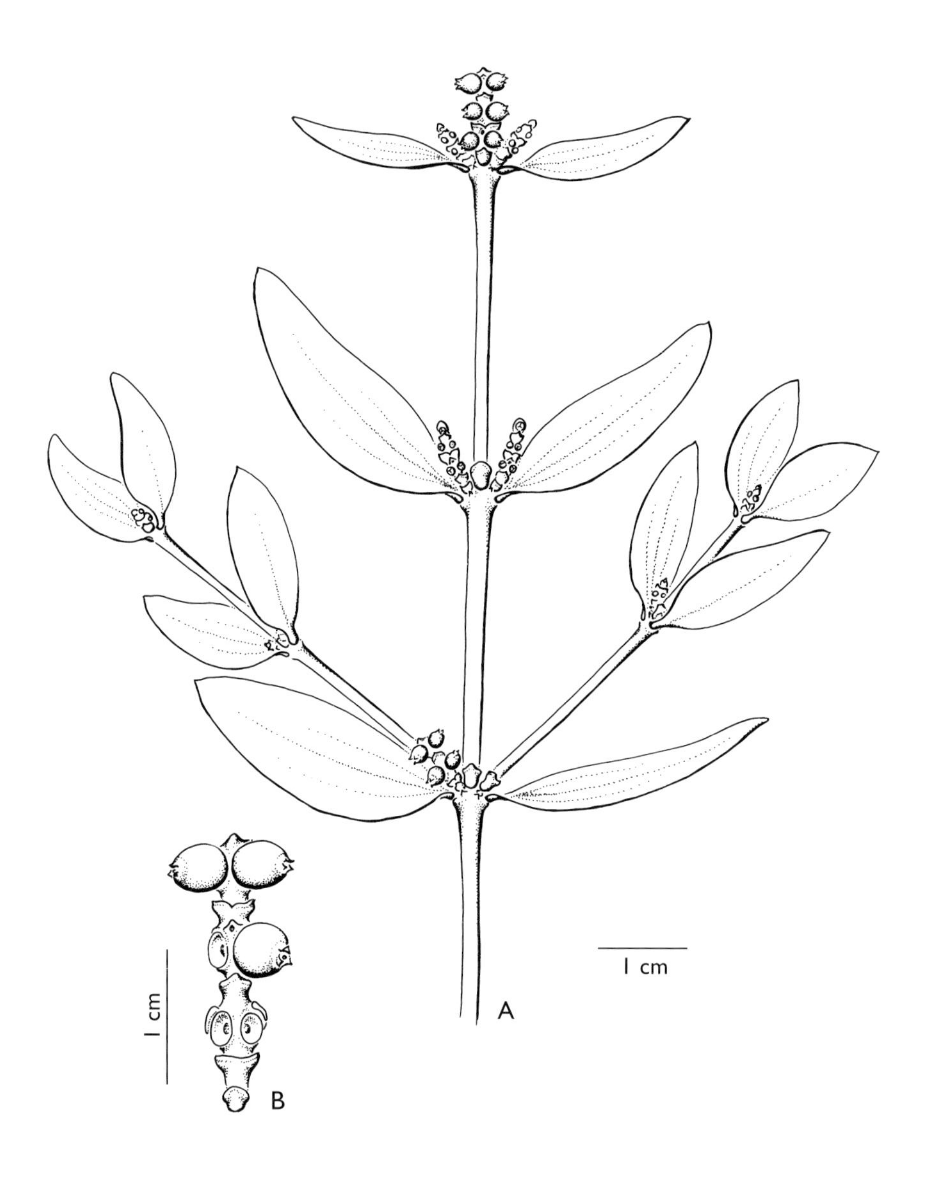

Fig. 50. *Phoradendron strongyloclados* Eichler: A, habit (the innovations more commonly have only one pair of foliage leaves); B, infructescence (reproduced from Kuijt, Syst. Bot. Monogr. 66, 2003).

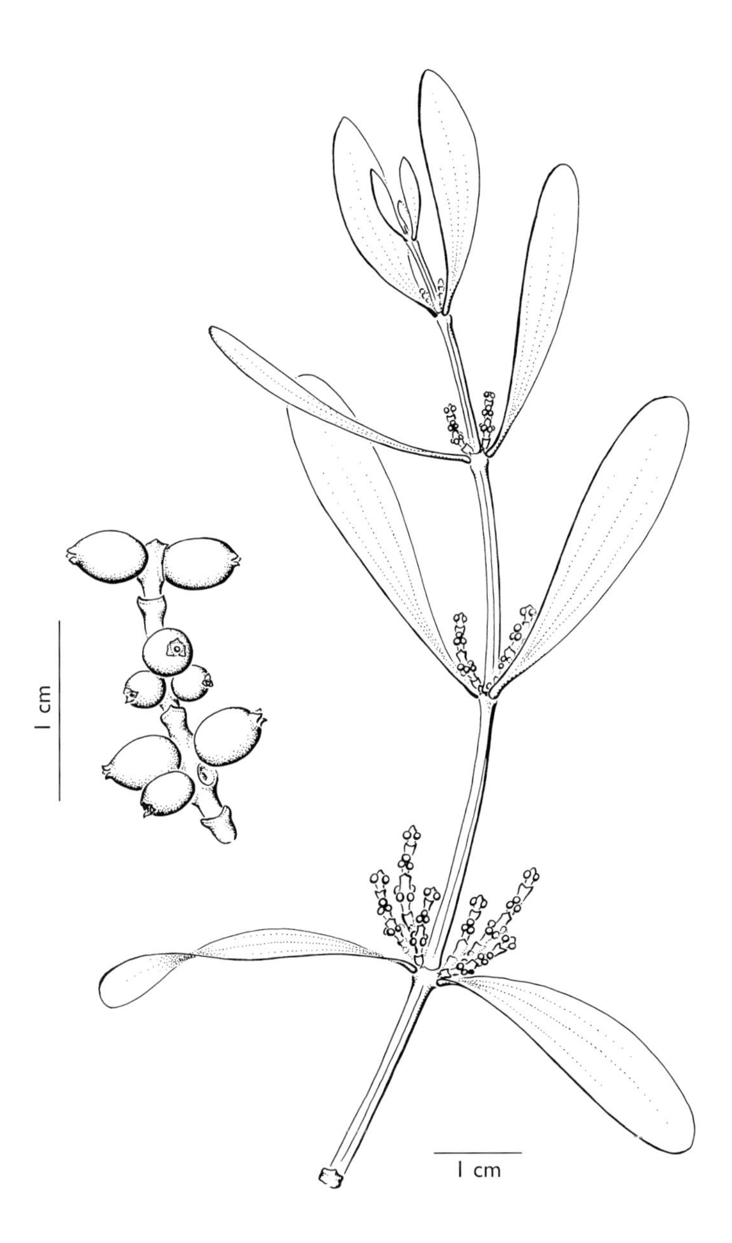

Fig. 51. *Phoradendron trinervium* (Lam.) Griseb.: habit and infructescence (reproduced from Kuijt, Syst. Bot. Monogr. 66, 2003).

Phoradendron appunii Trel., Phoradendron 104. 1916, as 'appuni'. Type: Guyana, Demerara, Appun 1783 (holotype K).
Phoradendron apertum Trel., Phoradendron 104. 1916. Type: Guyana, Demerara, Jenman 3801 (holotype K).
Phoradendron zuloagae Trel., Phoradendron 105. 1916. Type: Venezuela, Caracas, Zuloaga s.n. (not seen; photo in Trelease 1916: f. 9b, f. 150a).
Phoradendron theloneuron Rizzini, Rodriguésia 28(41): 26. 1976. Type: Venezuela, Bolívar, Wurdack & Monachino 41415 (holotype VEN, isotypes F, K, MO, P, RB, U, US).

Rather small, leafy plants; twigs quadrangular, becoming terete; internodes at first somewhat flattened to quadangular, to 6 cm long; basal cataphylls extremely inconspicuous, sometimes with a second pair 0.3-0.6 cm higher. Petiole not or scarcely distinguishable; blade rather thick, spatulate to elliptical, to 5 x 2 cm, base cuneate; usually 3 prominent palmate veins of about same length. Inflorescence to 2.5 cm long in fruit; peduncle simple, 0.1-0.4 cm long, or 0.5-0.7 cm long when subtended by a low pair of sterile bracts; fertile internodes 2 or 3, with 3-5 flowers per bract in 2-seriate pattern. Fruit dull orange, ovoid, relatively large, 0.5 x 0.3 cm, with rather narrow tip when dry, perianth segments more or less spreading or erect.

Distribution: Yucatan, Guatemala, Panama, the Caribbean, the Guianas, Venezuela, Colombia, Brazil; 300+ collections studied (GU: 9; SU: 5; FG: 4).

Selected specimens: Guyana: Kanuku Mts., Rupununi R., Bush Mouth near Witaru Falls, Jansen-Jacobs *et al.* 114 (K, LEA, MO, U, US); Rupununi Distr., Dadanawa, Rupununi R., Jansen-Jacobs *et al.* 2055 (F, K, LEA, MO, U, US); Crabwood Cr., along river, Jansen-Jacobs *et al.* 4353 (LEA, U); Soesdyke-Linden Hwy, savannas S of Timehri, International Airport, Kelloff *et al.* 616 (LEA). Suriname: Upper Coppename R., Boon 1157 (U); between Wia wia-bank and Grote Zwiebelzwamp, near km 8.8, Lanjouw & Lindeman 1238 (K, U); Saramacca R. headwaters, White Rock Rapids, below Posoegronoe, Maguire 24008 (F, GH, K, MO, U, UC, US); Wilhelmina Gebergte, hills 9 km N of Lucie and 12 km W of Oost Rs., Maguire *et al.* 54190 (RB); Herminadorp, Marowijne R., Wessels Boer 273 (F, GH, U, UC). French Guiana: estuary of Cr. Malmanoury, Billiet & Jadin 1729 (LEA); Mt. Chauve, Cremers *et al.* 14866 (CAY, LEA); Mts. Bakra, 2 km W of Pic Coudreau, de Granville 4160 (LEA, US); Camp No. 3, Roche No. 1, Akouba Booka goo Soula, Bassin du Ha, 500 m to SW, de Granville *et al.* 9770 (LEA, U, US).

Vernacular name: Suriname: koedjibjosore = kudibiu suru.

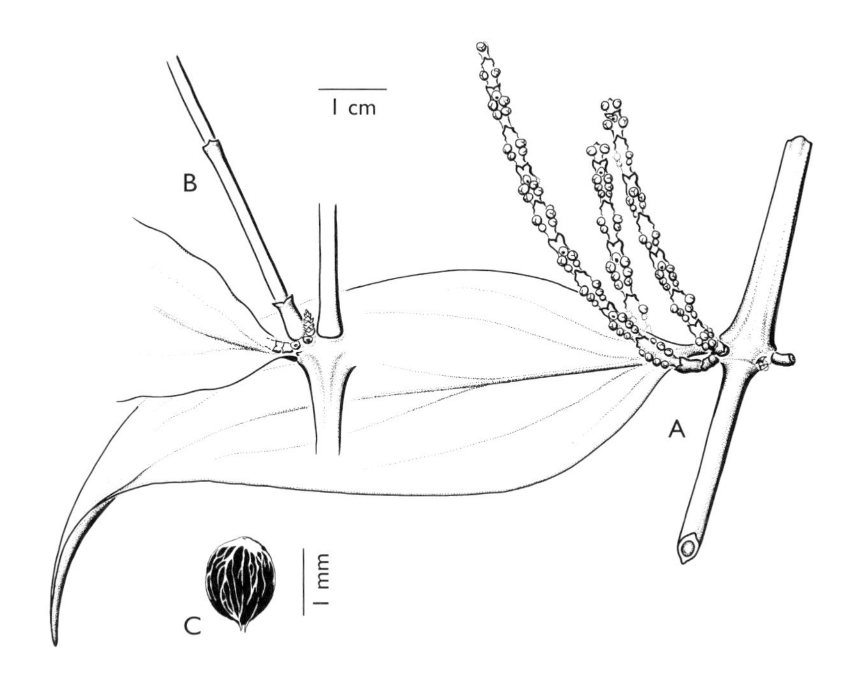

Fig. 52. *Phoradendron undulatum* (Pohl ex DC.) Eichler: A, node with leaf and inflorescences; B, detail of node showing basal cataphylls; C, seed (reproduced from Kuijt, Syst. Bot. Monogr. 66, 2003).

27. **Phoradendron undulatum** (Pohl ex DC.) Eichler in Mart., Fl. Bras. 5(2): 122. 1868. – *Viscum undulatum* Pohl ex DC., Prodr. 4: 282. 1830. Type: Brazil, Pohl s.n. (holotype W, destroyed, isotype G-DC). – Fig. 52

Phoradendron uleanum Steyerm. in Steyerm. *et al.*, Fieldiana, Bot. 28: 223. 1951. Type: Brazil, Roraima, Ule 8603 (holotype B, destroyed, see F Negative # 18191, isotype K).

Rather large species; internodes keeled and compressed especially below node, sometimes to 10-12 cm long; lowest lateral internodes terete; basal cataphylls 1-4 pairs, highest 2-5 cm above axil. Petiole ca. 1 cm long; blade lanceolate to narrowly ovate, to 17 x 7 cm, commonly half this size, apex acute; venation obscure but pinnate. Monoecious. Inflorescence axillary and at older nodes, each with 0-4 pairs of sterile cataphylls and 6-10 fertile internodes; flowers 5-7 per fertile bract, 2-seriate, terminal flower perhaps always male, others female. Fruit white, ovoid, 0.3 x 0.2 cm, perianth segments closed.

Distribution: Mexico (Chiapas), C America, the Caribbean, throughout tropical S America to amazonian Peru and Bolivia; no records have been seen from the Guianas and the Amazon Basin, but it will almost certainly be found there eventually, at least in western Guyana, it is known from adjacent Venezuela (Bolívar) and Brazil (Terr. Roraima).

TAXONOMIC AND NOMENCLATURAL CHANGES
NEW TYPIFICATIONS

Loranthaceae

New synonyms:
Cladocolea elliptica Kuijt to Cladocolea micrantha (Eichler) Kuijt
Phthirusa angulata K. Krause to Phthirusa stelis (L.) Kuijt
Psittacanthus corynocephalus Eichler to Psittacanthus acinarius (Mart.) Mart.
Psittacanthus collum-cygni Eichler to Psittacanthus eucalyptifolius (Kunth) G. Don

New typifications:
Phthirusa seitzii Krug & Urb.
Psittacanthus cucullaris (Lam.) Blume

Viscaceae

Lectotypification:
Viscum fockeanum Miq.

NUMERICAL LIST OF ACCEPTED TAXA

Eremolepidaceae

1. Antidaphne Poepp. & Endl.
   - 1-1. A. viscoidea Poepp. & Endl.

Loranthaceae

1. Cladocolea Tiegh.
   - 1-1. C. micrantha (Eichler) Kuijt
   - 1-2. C. nitida Kuijt
   - 1-3. C. sandwithii (Maguire) Kuijt

2. Gaiadendron G. Don
   - 2-1. G. punctatum (Ruiz & Pav.) G. Don

3. Oryctanthus (Griseb.) Eichler
   - 3-1. O. alveolatus (Kunth) Kuijt
   - 3-2. O. florulentus (Rich.) Tiegh.
   - 3-3. O. occidentalis (L.) Eichler subsp. continentalis Kuijt

4. Oryctina Tiegh.
   - 4-1. O. atrolineata Kuijt
   - 4-2. O. myrsinites (Eichler) Kuijt

5. Phthirusa Mart.
   - 5-1. P. coarctata A.C. Sm.
   - 5-2. P. disjectifolia (Rizzini) Kuijt
   - 5-3. P. guyanensis Eichler
   - 5-4. P. nitens (Mart.) Eichler
   - 5-5. P. podoptera (Cham. & Schltdl.) Kuijt
   - 5-6. P. pycnostachya Eichler
   - 5-7. P. pyrifolia (Kunth) Eichler
   - 5-8. P. rufa (Mart.) Eichler
   - 5-9. P. stelis (L.) Kuijt
   - 5-10. P. stenophylla Eichler
   - 5-11. P. trichodes Rizzini

6. Psittacanthus Mart.
   - 6-1. P. acinarius (Mart.) Mart.
   - 6-2. P. clusiifolius Eichler
   - 6-3. P. cordatus (Hoffmanns. ex Schult. & Schult. f.) Blume

6-4. P. cucullaris (Lam.) Blume
6-5. P. eucalyptifolius (Kunth) G. Don
6-6. P. grandifolius (Mart.) Mart.
6-7. P. lasianthus Sandwith
6-8. P. melinonii (Tiegh.) Engl.
6-9. P. peronopetalus Eichler
6-10. P. plagiophyllus Eichler
6-11. P. redactus Rizzini
6-12. P. robustus (Mart.) Mart.

7. Struthanthus Mart.
7-1. S. concinnus (Mart.) Mart.
7-2. S. dichotrianthus Eichler
7-3. S. gracilis (Gleason) Steyerm. & Maguire
7-4. S. syringifolius (Mart.) Mart.

8. Tripodanthus (Eichler) Tiegh.
8-1. T. acutifolius (Ruiz & Pav.) Tiegh.

Viscaceae

1. Dendrophthora Eichler
1-1. D. decipiens Kuijt
1-2. D. densifrons (Ule) Kuijt
1-3. D. elliptica (Gardner) Krug & Urb.
1-4. D. fanshawei (Maguire) Kuijt
1-5. D. perfurcata (Rizzini) Kuijt
1-6. D. roraimae (Oliv.) Ule
1-7. D. warmingii (Eichler) Kuijt

2. Phoradendron Nutt.
2-1. P. acuminatum Kuijt
2-2. P. aphyllum Steyerm.
2-3. P. bathyoryctum Eichler
2-4. P. bilineatum Urb.
2-5. P. chrysocladon A. Gray
2-6. P. crassifolium (Pohl ex DC.) Eichler
2-7. P. dipterum Eichler
2-8. P. granvillei Kuijt
2-9. P. hexastichum (DC.) Griseb.
2-10. P. inaequidentatum Rusby
2-11. P. krameri Kuijt
2-12. P. mairaryense Ule

2-13. P. morsicatum Rizzini
2-14. P. mucronatum (DC.) Krug & Urb.
2-15. P. northropiae Urb.
2-16. P. obtusissimum (Miq.) Eichler
2-17. P. pellucidulum Eichler
2-18. P. perrottetii (DC.) Eichler
2-19. P. piperoides (Kunth) Trel.
2-20. P. poeppigii (Tiegh.) Kuijt
2-21. P. pteroneuron Eichler
2-22. P. pulleanum K. Krause
2-23. P. quadrangulare (Kunth) Griseb.
2-24. P. racemosum (Aubl.) Krug & Urb.
2-25. P. strongyloclados Eichler
2-26. P. trinervium (Lam.) Griseb.
2-27. P. undulatum (Pohl ex DC.) Eichler

COLLECTIONS STUDIED
(Collection numbers in **bold** refer to types)

Eremolepidaceae

GUYANA

Pipoly, J.J., *et al.,* 10686 (1-1)

Loranthaceae

GUYANA

Acevedo, P., *et al.*, 3273B (5-2)
Anderson, 561 (5-8)
Archer, W.A., 2394 (3-2)
Boom, B.M., *et al.*, 9081 (2-1)
Clarke, D., *et al.*, 243 (5-8); 996 (7-3); 1650 (6-3); 3729 (5-11); 4400 (6-6); 4758 (5-7); 4819 (6-6); 5609, 5734 (2-1); 5764 (7-3); 6790 (1-1); 7029 (6-12); 7329 (1-1)
Cowan, R.S., 39254 (5-4); 39290 (5-8)
Cowan, R.S. & T.R. Soderstrom, 1726 (6-7); 1765 (5-8); 2103 (6-7)
Cremers, G., *et al.*, **10912** (4-1)
Cruz, J.S. de la, 1598 (5-9); 3072 (5-7); 3303 (5-9); 3661 (3-2); 3817 (5-9)
Davis, 624 (3-1); 2442 (5-8)
Drake, s.n. (5-8)
Ek, R.C., 748 (3-1); 763 (5-9)
Fanshawe, D.B., 2943, 6090 (3-2)
Forest Department, RB-37 (7-4); F-49, F-58, F-180 (5-8); D-275 (6-10); D-477 (5-8); D-585 (6-10); F-1088 (5-8); F-1783 (6-10); 2572 (3-2); 2690 (3-1); 2942 (5-6); 3080 (3-2); F-3454 (6-10); 3713 (3-1)
Gillespie, L.J., *et al.*, 788 (7-2); 929 (6-7); 949 (5-3); 1003 (7-2); 1139, 1177 (5-9); 1318 (5-2); 1328 (6-7); 1604, 2052 (5-9); 2203 (5-8); 2335 (3-1); 2690 (5-8); 2693 (7-3); 2876 (5-4); 2934 (5-8); 3055 (5-4); 3078 (5-2); 4288 (6-7)
Gleason, H.A., 417 (5-8); 920 (3-1)
Goodland, R., & R. Persaud, 686 (7-2)
Graham, E.H., 254 (5-1); 566 (3-2)
Hahn, W., *et al.*, 3918 (5-8); 4048 (6-7); 5143 (5-6); 5451 (2-1); 5572 (7-2); 5617, 5739 (5-9)
Harrison, S.G., 1467 (5-8)
Henkel, T.W., *et al.*, 42 (5-8); 653 (7-3); 762 (5-10); 1567, 2318 (6-7); 2544 (5-9); 3729 (6-3); 3807 (5-7); **4400** (1-2); 4772 (5-2); 5939 (6-7); 6017 (7-3)
Hill, R., *et al.*, 27201 (5-8)
Hitchcock, A.S., 16708 (7-2); 17111 (3-1); 17175 (3-2)
Hoffman, B., *et al.*, 710, 1621 (5-8); 1625 (7-3); 1675, 1778 (5-8); 1783, 1844 (6-7); 1845 (5-8); 1982 (7-3); 2518 (5-7); 3110, 3329 (6-7)
Huber, O. & S.S. Tillett, 5550 (5-3)
Im Thurn, E.F., s.n., (3-2); s.n. (7-4)
Irwin, H., *et al.*, 57500 (5-2)
Jansen-Jacobs, M.J., *et al.*, 99 (6-3); 1103 (7-2); 1990 (5-8); 2034 (6-3); 2599 (7-2); 3086 (5-9); 3087 (5-7); 3697 (1-1); 3908 (6-3); 4768 (6-3)

Jenman, G.S., 468 (3-1); 1241 (5-8); 1522 (3-2); 1680 (5-8); 2254 (3-1); 2529 (3-2); 2531 (7-4); 2534 (5-6); 2536 (3-2); 2538 (7-2); 2011 (3-2); **2531** (7-4); 2532 (3-1); 3635 (5-6); 3647, 3708, 3771 (7-2); 3789 (3-2); 3900 (7-4); 4122 (5-2); 4410 (7-2); 4895 (5-8); 5109, 5115 (7-4); 5343 (7-2); 5344 (5-6); 5345 (7-2); 5807 (3-2); 6639 (7-2); 6864 (3-1); 7395 (7-2); 7591 (7-2)
Kelloff, C.L., *et al.,* 902 (6-7)
Knapp, S. & J. Mallet, 2775, 2900 (5-8)
Kvist, L.P., *et al.,* 56 (6-7)
Lance, K., *et al.,* 66 (6-7)
Lanjouw, J. & J. van Donselaar, 930 (7-4)
Linder, D.H., 67, 131 (3-2)
Maas, P.J.M., *et al.*, 3531 (5-9); 3674, 3753 (7-2); 3953 (7-4); 4073 (6-3); 5438 (5-4); 5441, 5442 (7-2); 5806 (7-4); 7351 (6-12)
McDowell, T., *et al.*, 1871, 2580 (5-8); 2984 (7-3); 3135 (5-8); 3742 (5-9); 4171 (7-3); 4314 (5-7); 5806 (7-3)
Maguire, B., *et al.*, 22996 (3-1); 23077 (6-7); 23194 (5-8); 23278 (5-3); 23577 (7-2); 23578 (5-9); 32261 (6-7); 32620 (5-8); 32646, 40668 (6-7)
Mori, S.A., *et al.* 25307 (6-2)
Mutchnik, P., *et al.*, 161 (6-7); 1530 (5-8)
Parker, C.S., s.n. (3-2)
Persaud, 68 (3-1)
Pipoly, J.J., *et al.*, 7373 (5-8); 7388 (4-1); 7630, 7653 (6-7); 7759 (5-8); 7763 (6-7); 7899, 7946, 7985, 8419 (5-8); 8581 (7-3); 8820 (5-8); 8935 (5-2); 9052, 9152, 9335, 9616 (5-8); 9637 (5-9); 9681 (5-8); 9694 (5-9); 9811 (5-2); 9984 (6-7); 10024 (5-3); 10264, 10273 (7-3); 10363 (6-7); 10700 (6-9); 10735 (6-7); 10737 (5-8); 11104, 11175 (6-7); 11257 (7-2); 11277 (3-2); 11326 (5-2); 11417 (7-1)
Polak, M., *et al.,* 87 (3-2); 167 (3-1); 551 (5-6); 582 (5-3)
Pollard, G.R.M., 97 (7-2)
Sandwith, N.Y., 37 (7-2); **1404** (1-3); 201 (3-1); **313** (5-6); 331 (5-8); 421 (5-6); 683 (3-2); 689 (5-7); 1115 (5-6); 1143 (6-10); **1366** (6-7); 1370 (5-8); 1404 (1-3);1579 (3-2)
Schomburgk, Ri., **1602** (5-3)
Schomburgk, Ro., ser. I, 344 (6-4); ser. II, 764 = Ri. 1497a (7-4); ser. II, 855 = Ri. 1490B (5-7)
Smith, A.C., 2114 (3-1); **2204** (5-1); 2232, 2261, 2322 (7-2); 2325 (5-9); 2385 (3-2); 2386 (5-7); 2435 (7-4); 2769 (5-7); 2849 (5-2)
Stockdale, F.A., 8839 (5-7)
Stoffers, A.L., *et al.*, 148 (5-8); 482 (5-7)
Tate, G.H.H., 49 (7-2)
Tillett, S.S., *et al.*, 43993, 45104 (6-7); 45311 (5-8); 45378 (6-7); 45878 (6-2)
Tutin, T.G., 21 (3-1); 205 (7-4); 297 (5-8); 443 (7-2)

SURINAME

Archer, W.A., 2791 (3-2)

FRENCH GUIANA

Cremers, G., *et al.*, 982 (6-8); **9424** (5-6); 9872 (6-8); 9899 (6-6); 11238 (5-9); 12770 (5-6); 12937 (5-7); 14867 (5-6); 15291 (6-2); s.n. (5-7)
Descoing, B., & C. Luu, 20038 (5-7); 20291 (5-9); 20299 (5-7)
Feuillet, C., *et al.*, 3594 (6-6); 9966 (6-1)
Fournet, A., 142 (5-7)
Gabriel, s.n. (3-2)
Gely, A., 135 (5-7)
Granville, J.J. de, *et al.*, 2076 (5-7); 3228 (5-6); 4086, 4113 (6-11); 9330 (5-7); 9608 (4-2)
Grenand, P., 1470 (6-4)
Hijman, M. & J. Weerdenburg, 200 (7-2)
Hoff, M., *et al.*, 5088, 5373, 6587 (5-7)
Jacquemin, H., 2119 (6-6)
Larpin, D., 489 (6-11), 870 (5-9)
Leblond, J.B., **222** (3-2)
Leeuwenberg, A.J.M., 11755 (5-9)
Lemée, A.M.V., s.n. (3-1); s.n. (3-2); s.n. (5-7); s.n. (5-9); s.n. (6-10)
Leprieur, F.R.M., **s.n.** (3-2); s.n. (6-4)
Martin, J., s.n. (5-7); s.n. 7-4)
Mélinon, E., 19 (5-9); 21 (3-2); 29 (5-7); **145** (6-8); 197 (5-7); 216, 320, 426, 1100 (5-9); s.n. (5-7); s.n. (6-10)
Mori, S.A., *et al.*, 18118 (7-3); 18185 (7-1); 18896 (6-1); 23848 (7-3); 23969 (5-6); 25215 (6-6); 25307 (6-2, see note to 6); 25495 (6-11)
Moricand, 210 (3-1)
Oldeman, R.A.A., *et al.*, B-933, B-973, 983, 1090 (5-9); B-1486 (6-9); 1675, 1676, 1802 (5-9); 2205 (6-4); 3441, B-3913 (5-9)
Perrottet, G.S., **227**, s.n. (3-2); s.n. (5-6); **s.n.** (5-7)
Poiteau, P.A., s.n. (3-2); s.n. (3-3); **s.n.** (5-6); s.n. (5-7)
Prance, G.T., *et al.*, 30630 (5-6)
Prévost, M. F., *et al.*, 338 (7-3); 997 (6-4); 1360 (6-6); 1959 (5-7); 3408 (1-1)
Roubik, D., 237 (5-9)
Sagot, P.A., 297 (3-2); 298 (5-9)
Sarthou, C., 262 (6-11)
Sastre, C., *et al.*, 280 (5-9); 3828 (5-7); 3916 (5-7); 8010 (5-9); 8177 (5-9)
Sauvain, M., 468 (5-6)
Schnell, R., 11226, 11236, 11386, 11387, 11389 (5-9); 11970 (5-7)
Wurdack, J., 4045 (3-1)

Viscaceae

GUYANA

Acevedo, P., *et al.*, 3403, 3494 (2-18); 6046 (2-6)
Allison, 149 (2-15); 150 (2-25)
Andel, T.R. van, *et al.*, 1757 (2-18)
Appun, C.F., **1783** (2-26)
Archer, W.A., 2262 (2-19)
Bailey, I.W., 132 (2-18)
Bartlett, A.W., 8439 (2-19)
Boom, B.M., *et al.*, 7129 (1-4); 7138 (2-25); 8296 (2-6); 9301 (2-21)
Christenson, E., *et al.*,1875 (2-4)
Clarke, D., *et al.*, 808 (2-6); 862 (2-17); 1127 (2-19); 1159 (2-4); 1410 (2-19); 1780 (2-6); 1851 (2-19); 2974 (2-25);

SURINAME

Lanjouw, J. & J.C. Lindeman, 101 (2-25); 338 (2-19); 1238 (2-26); 1416 (2-19); 1466 (2-11); 2825 (2-6); 3473 (2-4)
Lindeman, J.C., 3690 (2-19); 5575 (2-19); 6266 (2-4)
Lindeman, J.C., A.L. Stoffers *et al.*, 812 (2-22)
Maas, P.J.M. & J.W. Tawjoeran, 3351 (2-6
Maguire, B., *et al.*, 23974 (2-6); 24008 (2-26); 24655 (2-24); **54147** (2-16); 54190 (2-26)
Mori, S.A., *et al.*, 8402 (2-19)
Oldenburger, F.H.F., *et al.*, 814 (2-21)
Pulle, A.A., 359 (2-18)
Samuels, J.A., 114 (2-18); 114 p.p. (2-19); 451 (2-7); 458 (2-19)
Sauvain, M., 217 (2-24)
Schulz, J.P., 7266 (2-18); 10369a (1-4)
Splitgerber, F.L., 240 (2-7); 241 (2-19); 816 (2-16)
Stahel, G., 173 (2-24), s.n. (2-25)
Teunissen, P.A. & J.Th. Wildschut, LBB, 11767 (2-6)
Versteeg, G.M., **239** (2-25); 241 (6-1); 901 (2-16)
Wessels Boer, J.G., 273 (2-26); 704 (2-19); 1120 (2-3)

FRENCH GUIANA

Acevedo, P., 4883 (2-6)
Billiet, F. & B. Jadin, 1245 (2-6); 1729 (2-26); 1732 (2-18); 6360 (2-19)
Broadway, W.E., 612, 722 (2-25)
Cremers, G., *et al.*, 5061 (2-25); 5552 (2-19); 6020 (1-7); 7213, 9552, 9904 (2-19); 13933, 13937 (2-4); 14258, 14372, 14458 (2-19); 14865 (2-6); 14866 (2-26); 15288 (2-6)
Fleury, M., 433 (2-19).
Foresta, H. de, 460 (2-19)
Gentry, A., *et al.*, 63182 (2-24)
Granville, J.J. de, *et al.*, 96 (2-19); 4160 (2-26); 4399 (2-6); 6927 (2-19); 8026 (2-10); 8168 (2-24); 9770 (2-26); 9933 (2-6); 10015 (2-24); 10077 (2-19); 10529 (2-22); **10802** (2-8); 10858, 12024 (2-19)
Grenand, P., 541 (2-18)
Hallé, F., 582 (2-24)
Hoff, M., 6754 (2-19)
Jacquemin, H., 2592, 2720 (2-19)
Leeuwenberg, A.J.M., 11788 (2-19)
Maas, P.J.M., *et al.*, 2209 (2-18)
Mori, S.A., *et al.*, 14949, 18090 (2-6); 21077, 22130 (2-19); 22778 (2-6); 23287 (2-24); 23977 (2-6)
Oldeman, R.A.A., *et al.*, 204, B-1085, 1370 (2-19)
Patris, J.B., s.n. (2-19)
Perrottet, G.S., **228** (2-18)
Prance, G.T., *et al.*, 30626 (2-18)
Prévost, M.F., 3396 (2-19)
Richard, L.C., **s.n.,** (2-24)
Sagot, P.A., 296 (2-19)
Sauvain, M., 371 (2-18)
Stoupy, s.n. (2-24)

INDEX TO SYNONYMS, NAMES IN NOTES AND SOME TYPES

Viscaceae

INDEX TO VERNACULAR NAMES

Viscaceae

Alphabetic list of families of series A occurring in the Guianas

Defined as in Cronquist, 1981, and numbered in his sequence, with alternative names. Those published, with chronological fascicle number and year.

Family	No.	Fascicle
Abolbodaceae		
(see Xyridaceae	182)	15. 1994
Acanthaceae	156	23. 2005
(incl. Thunbergiaceae)		
(excl. Mendonciaceae	159)	
Achatocarpaceae	028	22. 2003
Agavaceae	202	
Aizoaceae	030	22. 2003
(excl. Molluginaceae	036)	22. 2003
Alismataceae	168	
Amaranthaceae	033	22. 2003
Amaryllidaceae		
(see Liliaceae	199)	
Anacardiaceae	129	19. 1997
Anisophylleaceae	082	
Annonaceae	002	
Apiaceae	137	
Apocynaceae	140	
Aquifoliaceae	111	
Araceae	178	
Araliaceae	136	
Arecaceae	175	
Aristolochiaceae	010	20. 1998
Asclepiadaceae	141	
Asteraceae	166	
Avicenniaceae		
(see Verbenaceae	148)	4. 1988
Balanophoraceae	107	14. 1993
Basellaceae	035	22. 2003
Bataceae	070	
Begoniaceae	065	
Berberidaceae	016	
Bignoniaceae	158	
Bixaceae	059	
(incl. Cochlospermaceae)		
Bombacaceae	051	
Bonnetiaceae		
(see Theaceae	043)	
Boraginaceae	147	
Brassicaceae	068	
Bromeliaceae	189	p.p. 3. 1987
Burmanniaceae	206	6. 1989
Burseraceae	128	
Butomaceae		
(see Limnocharitaceae	167)	
Byttneriaceae		
(see Sterculiaceae	050)	
Cabombaceae	013	
Cactaceae	031	18. 1997
Caesalpiniaceae	088	p.p. 7. 1989
Callitrichaceae	150	
Campanulaceae	162	
(incl. Lobeliaceae)		
Cannaceae	195	1. 1985
Canellaceae	004	
Capparaceae	067	
Caprifoliaceae	164	
Caricaceae	063	
Caryocaraceae	042	
Caryophyllaceae	037	22. 2003
Casuarinaceae	026	11. 1992
Cecropiaceae	022	11. 1992
Celastraceae	109	
Ceratophyllaceae	014	
Chenopodiaceae	032	22. 2003
Chloranthaceae	008	24. 2007
Chrysobalanaceae	085	2. 1986
Clethraceae	072	
Clusiaceae	047	
(incl. Hypericaceae)		
Cochlospermaceae		
(see Bixaceae	059)	
Combretaceae	100	
Commelinaceae	180	
Compositae		
(= Asteraceae	166)	
Connaraceae	081	
Convolvulaceae	143	
(excl. Cuscutaceae	144)	
Costaceae	194	1. 1985
Crassulaceae	083	
Cruciferae		
(= Brassicaceae	068)	
Cucurbitaceae	064	
Cunoniaceae	081a	
Cuscutaceae	144	
Cycadaceae	208	9. 1991
Cyclanthaceae	176	
Cyperaceae	186	
Cyrillaceae	071	
Dichapetalaceae	113	
Dilleniaceae	040	
Dioscoreaceae	205	
Dipterocarpaceae	041a	17. 1995
Droseraceae	055	22. 2003
Ebenaceae	075	
Elaeocarpaceae	048	
Elatinaceae	046	

Plumbaginaceae	039	
Poaceae	187	8. 1990
Podocarpaceae	211	9. 1991
Podostemaceae	091	
Polygalaceae	125	
Polygonaceae	038	
Pontederiaceae	197	15. 1994
Portulacaceae	034	22. 2003
Potamogetonaceae	171	
Proteaceae	090	
Punicaceae	097	
Quiinaceae	045	
Rafflesiaceae	108	
Ranunculaceae	015	
Rapateaceae	181	
Rhabdodendraceae	086	
Rhamnaceae	116	
Rhizophoraceae	101	
Rosaceae	084	
Rubiaceae	163	
(incl. Henriquesiaceae)		
Ruppiaceae	172	
Rutaceae	132	
Sabiaceae	018	
Santalaceae	104	
Sapindaceae	127	
Sapotaceae	074	
Sarraceniaceae	054	22. 2003
Scrophulariaceae	153	
Simaroubaceae	130	
Smilacaceae	204	
Solanaceae	142	
Sphenocleaceae	161	
Sterculiaceae	050	
(incl. Byttneriaceae)		
Strelitziaceae	190	1. 1985
Styracaceae	076	
Suraniaceae	086a	
Symplocaceae	078	
Taccaceae	203	
Tepuianthaceae	114	
Theaceae	043	
(incl. Bonnetiaceae)		
Theophrastaceae	079	
Thunbergiaceae		
(see Acanthaceae	156)	
Thurniaceae	185	
Thymeleaceae	095	
Tiliaceae	049	17. 1995
Trigoniaceae	124	21. 1998
Triuridaceae	174	5. 1989
Tropaeolaceae	135	
Turneraceae	061	
Typhaceae	188	
Ulmaceae	020	11. 1992
Umbelliferae		
Urticaceae	023	11. 1992
Valerianaceae	165	
Velloziaceae	201	
Verbenaceae	148	4. 1988
(incl. Avicenniaceae)		
Violaceae	060	
Viscaceae	106	25. 2007
Vitaceae	117	
Vochysiaceae	123	21.1998
Winteraceae	001	
Xyridaceae	182	15. 1994
(incl. Albolbodaceae)		
Zamiaceae	208a	9. 1991
Zingiberaceae	193	1. 1985
(excl. Costaceae	194)	
Zygophyllaceae	133